Christian Science and Christ and Christmas

by Rolf A. F. Witzsche

version 2

Exploring
The Acme
Of Christian Science

This book is dedicated in honor of Mary Baker Eddy, one of the greatest scientific geniuses of her time. Her stature as a world-historic person reflects a number of pioneering achievements that she is widely honored for, which have uplifted the face of civilization in numerous respects. She is noted for her dedication to advancing the status of women; for ennobling Christianity with a scientific dimension; for reinstating its nearly lost element of Christ-healing; for her discovering and founding of Christian Science, and the writing of its textbook; for establishing a world-wide church and ordaining for it two impersonal pastors, the Bible and her textbook; and not least for her founding of the international newspaper, The Christian Science Monitor. However, these types of biographies rarely focus on her accomplishments as a pioneer in the scientific dimension where she was so far advanced of her time that she noted that "Future ages must declare what the pioneer has accomplished." These words remained on the first page of the preface of her textbook on Christian Science, Science and Health with Key to the Scriptures, which she had constantly revised and upgraded in the course of more than 400 editions of the book, keeping the book in line with her own scientific development over the space of 35 years since its first publication.

In celebrating the150th Anniversary of her discovery of Christian Science, it is appropriate that the historic dimension of Mary Baker Eddy be enriched with a specific focus on the scientific aspects of her work. This is the purpose of this book. She has accomplished enormously on this front, far more than what society gives her credit for, most of which remains generally unknown. If a historian had asked her in her time what she had most wished for in her life, she might have answered that she had labored for 44 years to give to humanity the clearest sense possible of the "acme" of Christian Science., which means that all other accomplishments where subsequent to it and reflect it. But what was her sense of the "acme" of Christian Science as she used the phrase at the end of the last chapter of her textbook? This answer was not provided. She gave hints of how it may be discovered, since Christian Science was a discovery from the beginning. She left the answer for future ages to declare, when society was sufficiently advanced to take note of the scientific dimensions that she had incorporated into her work, by which they would become discovered and be advanced.

Contents

The Acme of Christian Science

Mary Baker Eddy, the discoverer and founder of Christian Science, refers extensively to St. John's vision of a 4-square city coming down from God to humanity*1. and writes that **"his vision is the acme of this Science."***2

If this is her perception, I want to explore it. I want to explore what her sense of the acme is, of the greatest spiritual science of our time. Wouldn't anyone? So, let us begin.

Mary Baker Eddy describes Johns vision in great detail. She provides definitions for it in the last chapter of her textbook on Christian Science, *Science and Health with Key to the Scriptures,* which she is the author of. It is there, at the end of the last chapter, as if it was a summation of her life-work, that she writes under the last subheading, *The city of our God*, the following paragraph.

> "The writer's present feeble sense of Christian Science closes with St. John's Revelation as recorded by the great apostle, for **his vision is the acme of this Science** as the Bible reveals it. *2

In his vision, John, the Revelator, saw the great city as a profound scientific construct with such a far-reaching universal significance that he wrote, "the kings of the earth do bring their glory and honour into it." *3.

Mary Baker Eddy didn't leave John's vision standing as a mere promise. She fulfilled the promise in as much as anyone could. She didn't leave John's vision as a faceless, featureless construct. When she was done, she had brought all of her works into it, by which it was defined, and by which her work itself was defined in return. She summed it up with one word. She used the word "acme," the "acme" of all that she stood for, that she named "Christian Science". Nor did she merely describe John's vision of the city extensively in terms of its four-fold dimensions, and provided scientific definitions for it, but brought 8 major structures of her own creation into it that are monumental by themselves, but become vastly greater in their scientific significance by them all working together as a combined 'functional' unit.

The 8 major structures that Mary Baker Eddy had built onto the ACME foundation, are of two distinct types. Here it gets interesting. She created two types of structures. She created development oriented structures, and platform-type structures.

*1 Revelation 21:16 -
*2 textbook p.577:28 -
*3 Revelation 21:24

Developmental (progressive-type) structures	Platform-type structures
The Christian Science textbook - 16 chapters **Christ and Christmas** – 16 verses **The Lord's Prayer** - 16 stanzas **The Church Manual** – 16 segments	**The Platform of Christian Science** – 32 elements the counter platform '**mortal mind**' – 16 parts The chapter, **Recapitulation** – 24 elements The **Bible Lessons** – 26 topics ....the counter platform - **Adam** – 26 parts Note: the 26 parts reflect the weeks in 1/2 year **The Glossary definitions** – 144 definitions (9x16)

The development-focused structures characteristically begin their specific exploration of what is real by starting with the vaguest concept associated with the lowest point in the first column.

From there the progression sequences upwards to the highest point in a column, and onwards, repeated column by column.

The platform structures, in contrast, begin with the highest concept at the highest point, or the ultimate point, and sequence downward to what lies below. For two of these platforms, Mary Baker Eddy has created a counter platform each, a platform of related errors. And beyond these, Mary Baker Eddy created an advanced type of platform that defines the ACME structure in minute details for which she provided the building blocks, but placed the building of it into the courts of individual scientific determination.

.

This super-platform is the 144-part **platform of the Glossary definitions.**

So let's begin with the two basic types. These are structures for which Mary Baker Eddy provided both the building blocks and a specific sequence. Only the type remains to be determined, which determines how the structures apply to the ACME foundation. The resulting different sequences are shown below for the two basic types

Development progression				Platform statement sequence			
4	8	12	16	4	3	2	1
3	7	11	15	8	7	6	5
2	6	10	14	12	11	10	9
1	5	9	13	16	15	14	13

Since the 16-element ACME Structure is two-dimensional, and the application of the associated parts depends on their type, I have created a universal set of two-dimensional reference notations that refer to a specific location within the 4-square or ACME structure. I have applied the notations throughout this book. They indicate which column, and which level within a column, a specific element is located at.

Acme reference notations	#ooo	o#oo	oo#o	ooo#
Level 1	1ooo	o1oo	oo1o	ooo1
Level 2	2ooo	o2oo	oo2o	ooo2
Level 3	3ooo	o3oo	oo3o	ooo3
Level 4	4ooo	o4oo	oo4o	ooo4

The Acme of Christian Science: Its Infinite Destiny

Before the 4-square city was seen on the horizon
John the Revelator had seen a great wonder in heaven;
a woman clothed with the sun, and the moon under
her feet, and upon her head a crown of twelve stars.

The woman is humanity, and the 12 stars in her crown prefigure the factor 12 that John had used in part to describe the city foursquare. While Mary Baker Eddy had referred to the 'city' as the acme of Christian Science, she had made no reference in her description of the 'city' to the factor 12 as John had done. Is this a paradox? No, it isn't. Mary Baker Eddy did more than merely refer to the factor 12 in her defining the 'city', she made it one of its most profound features, a feature that is critical for the future of humanity, including whether humanity will have a future.

The factor 12 is incorporated into the structural division of the 4-square city into a top row that reflects functionally what Mary Baker Eddy has termed, the SCIENTIFIC TRANSLATION OF IMMORTAL MIND, and three lower rows that combine 12 elements that reflect together, functionally, what she termed the SCIENTIFIC TRANSLATION OF MORTAL MIND.

SCIENTIFIC TRANSLATION OF **IMMORTAL MIND**
GOD / MAN / IDEA

SCIENTIFIC TRANSLATION OF **MORTAL MIND**
Third Degree: Spiritual / **Understanding / Reality**
Second Degree: Evil beliefs disappearing / **Transitional qualities / Moral**
First Degree: Depravity **/ Unreality / Physical**

With her major division of the ACME structure into two parts, divided by the concept of IMMORTAL MIND and MORTAL MIND, she raised one of the most puzzling questions that is universally critical, The question is: **What is mortal mind?**

In Christian Science, the very concept of a mortal mind is a contradiction in itself, because Mind is immortal and cannot be reflected in mortality. Mortality applies only to false concepts that have their end built into their natures as being false. When the true is understood, the false vanishes from sight as if it had never existed. Until then, the illusions weigh heavily on the mind and shape human experience.

So, what is mortal mind then, as a concept?

The answer is simple. It is not a counter-mind, but is merely a feeble sense of Immortal Mind expressed in man and humanity. Mortal mind can be seen functionally as a mental fog that results in the error of small-minded thinking that in the extreme can have 'mortal' consequences. Small-minded thinking is never without consequences. Depravity, for example is the outcome of self-deprivation. It has consequences.

Mortal mind can be likened to a traveler who aims to explore the world with the eyes closed; who cannot even see himself. Then the Christ comes along and says to the traveler "You do not have to live in this mode of severe self-deprivation. Open your eyes, man." He points to the first row and says to the traveler, "You are Spirit, because God is Spirit. God and man is one. The difference is a difference in office, but not in being. And because you are Spirit, everything about you is Spirit-like, baring the attributes of Spirit, like beauty, health, harmony, completeness, perfection; this renders you spiritual through and through in every respect; and every person in the world so, likewise. If you ever are in doubt about who and what you are, look at the first row. You will find yourself there. The synonyms for God, are attributes of you. The attributes are expressed in your being, spiritually, divinely, intelligently, truthfully, and so on. When you see yourself in this manner, your small-minded notions will fall away as if they never existed, because the world is small only in dreams with the eyes closed. So, open your eyes as wide as you can, and then some more."

Mary Baker Eddy has defined 3 stages of severity of small-mindedness.

The most extreme severity of small-mindedness, Mary Baker Eddy terms "depravity." Depravity is the result of deprivation. In spiritual terms, society deprives itself when its vision is too narrow for the truth to be seen. For example, when society becomes encumbered with materialistic concepts and conventions, it cannot see in this forest of trash that the world is inherently wide and profound and man is divine, because God is divine. In this dense forest of trash, the self-deprivation is extreme and the result is typically fascist in nature, which is a type of total void of humanity.

Whenever a society looses its grounding in the divine and fails to recognize its divinity, and thereby fails to recognize everyone else's divinity by virtue of the all-ness of God, then the result of the ensuing grand deprivation becomes utter chaos that is typically immensely destructive. The universality of God renders, love, necessarily universal in nature. Anything less than universal love is not legitimate, but is self-deprivation, and dangerously so.

World War II stands as an example. It resulted from near universal small-minded perceptions, instead of living universal love. While the small-minded thinking in this case was politically imposed, which a small-minded society had failed to develop the inner resources to resist and defend itself against, being ignorant of the truth, the effect that became war, was nothing more than the outcome that is typical for universal self-deprivation. As a consequence of this fundamental failure in society, to understand and acknowledge its humanity as a precious gem, and as the most valuable asset is has, large segments of the Eurasian continent were demolished, and

more than 50 million people ended up murdering one-another in various orgies of brutality.

Christian Science offers an escape path, which may be the only escape path that humanity has. Christian science brings scientific light into the dark places. It brings a sense of reality back into the scene. Christian Science is thereby uniquely qualified to free society from its small-minded trap. In recognition of this capability, Mary Baker Eddy has termed the bottom row, the domain of Christian Science:
> "Christian Science, which today and forever interprets the great example and the great Exemplar." (S&H 577:18)

Christian Science extends a hand to those who have trapped themselves into the sewer, saying in a scientific manner, "You don't have to be there. This is not your native home. Self-deprivation is not a legitimate way of living. There is a way out."

That's where the ACME begins to unfold. We need this unfolding of the ACME of Christian Science badly in the modern world, because much of the world is still stuck deep in the mortal-mind trap of small-minded thinking. We face a horizon of looming tragedies, such as large-scale economic and financial collapse, new forms of fascism, relentless terrorism, endless wars, and a final nuclear war that no one may survive, which is fully prepared for. And in addition to all that, we face the impending collapse of humanity resulting from the phase shift of the solar system towards the next Ice Age to be starting in the 2050s that very few will be able to live past if the preparations for it are not being made, which are not even on the agenda today.

These are all cases of small-minded thinking in the extreme, where the self-deprivation of society is great, and dangerously extensive. There exist several degrees of healing that can get society out of its trap.

The first degree in healing will get us out of this fascist trap and to higher ground, which Mary Baker Eddy has termed "moral." She has defined the higher ground above the hell-hole of depravity as the domain of Christianity:
> "Christianity, which is the outcome of the divine idea in Christian history." (S&H 577:16)

At this higher ground, however, society is not fully

secure yet. The world's moral concepts have still many hidden elements of depravity attached. While visions of truth are clearer there, small-minded thinking nevertheless continues and rules the day. This means that a second degree in healing needs to be accomplished to get society out of the doldrums. We need to get to higher ground, because the moral ground is not essentially spiritually true.

Sometimes society's sense of morality is outright false, in the scientific sense. For example, it is deemed moral today to massively burn food in a starving world, by converting it to biofuels that are burned in cars. This modern travesty is so massive that the food that is burned would normally nourish 400 million people. The burning of this massive amount of food likely murders 100 million people per year with starvation. This ongoing genocide is deemed moral by society as everyone participates in the process at the gas pump. We need to get to still-higher ground just to rescue ourselves from this folly.

The second degree of healing that is natural here, gets one to the second row that Mary Baker Eddy defined as the Christ:

> "The Christ, the spiritual idea of God." (S&H 577:15)

Here we have justified hope at the second level. There, small-minded thinking has diminished by two degrees. In the light of this progress we begin to see man as divine idea, and we begin to love what we see.

The term Christ is an ancient concept. In the scientific sense, though, the Christ idea is not mysterious and far off. It is practical, immediate, and powerful. It simply signifies that the divinely normal process of divine **Love reflected in love**, is happening. By the unfolding of this process in thought and in deed, we begin to experience the real nature of the universe, the divine reality, evermore and everywhere, as we invariably begin to love what we see as the greatest gem in the universe, and begin to act accordingly.

On this level, on the level of the Christ where the wonders of Love become the focus, reflected in universal love for our divinity, a third degree of healing becomes possible and normal. In the flow of this healing the last vestiges of small-minded thinking fall by the wayside. The total healing of small-minded thinking is inevitable at the Christ-level, because the divine Truth that comes to light is not trivial and small, but is huge and profound. In this profundity we finally find our divine identity.

By the third degree of healing, war, nuclear war, all terror and genocide, and so forth, fall by the wayside. In the process of the third degree healing, we will prepare our world without fail for the coming Ice Age phase shift that will be upon us in the 2050 according to a large body of clear physical evidence that we can intelligently trust with great certainty. With the third-degree healing we will begin to built the needed 6,000 new cities for a million people each, to enable the nations outside the tropics, to relocate themselves into the tropics before the Ice Age phase shift begins and their present territories become uninhabitable. With small-minded thinking out of the picture, we will complete the job in 30 years as we must.

There is nothing trivial about the process of healing of the third degree, by which the divine is embraced humanly, because the divine is huge, profound, powerful. The third-degree healing takes us far beyond the dark concepts of small-minded thinking that thereby are simply left behind as if they never existed. Then life is beautiful, rich and amazing, as it naturally is in the divine.

The ACME foundation enables this type of progressive development to happen, which nothing less can achieve. When Mary Baker Eddy began to build on the ACME foundation, she evidently realized with joy at every step along the way that this is absolutely the ACME of Christian Science where movements can become possible that we can't even dream of as yet. In comparison with that, she may have realized that even her great, revolutionary understanding of Christian Science was still feeble in comparison with what can yet be achieved.

Mary Baker Eddy's definition of the 4-levels

This section presents a summary of the 4 definitions that Mary Baker Eddy has provided for the 4 rows of the ACME structure, which all, together, evidently play a role by their relationships to healing.

"**The Word** of Life, Truth, and Love (**omnipotence**) **Truth** HEAVEN – DAY	**God / Man / Idea** (Man as God's image is seen)
"**The Christ**, the spiritual idea of God." (**omniscience**) **Love** KINGDOM OF HEAVEN – MORNING	**Third degree**: understanding (spiritual)
"**Christianity**, which is the outcome of the divine idea in Christian history." (**omnipresence**) EARTH – EVENING	**Second degree**: transitional (moral)
"**Christian Science**, which today and forever interprets the great example and the great Exemplar." (**omni-action**) HELL – NIGHT	**First degree**: depravity (physical)

Note: The first definition is from the chapter The Apocalypse. The definition in brackets, is from the Glossary definition of the term Good. The terms, Truth, and Love on the top two rows, are the sequential parts of the terms for God that are not specifically defined in the Glossary as they have a larger significance. See the section in this book, "Soul, Truth, and Love".

The terms shown in all upper case letters are two sets of four applicable terms that are located in the Glossary, which are of a type that represents distinct levels. Heaven, Kingdom of heaven, Earth, and Hell are one set. Day, Morning, Evening, and Night are another set.

Where do we go from here? Eternity?

Eternity is the destiny of the truth, but not an eternity in time, rather, in terms of the ultimate and absolute that exists as truth today as it has always existed and will forever exist, because God is Truth. We can find this out and experience its essence, because Truth is true. Thus, we, humanity, have the capacity to know the truth and to experience it. While the light of divine Truth may be hidden for a season in such landscapes where the great moments in history find society, "a 'little' people," - a people made small by cultivated smallness in thinking - the potential nevertheless exists, and has always existed, for us all, individually, to reach for the truth from the foundation of St. John's science, which Mary Baker Eddy has referred to as the ACME of Christian Science.

Mary Baker Eddy's TRANSLATION OF IMMORTAL MIND presents the eternal and absolute. Those who dwell in this light become 'immortal' as the Son of God by their actions, because this is the nature of man and the quality of man's being. Being is not immortal in time, but immortal in eternity without time. A touch of Truth defines eternity, and its result uplifts the world forever. Christ Jesus had this effect. Mary Baker Eddy had this effect. The next step on the ladder is furnished by the ACME of Christian Science itself.

Because the two types of translations, the

TRANSLATIONS OF IMMORTAL MIND and the TRANSLATION OF MORTAL MIND are integral parts of the ACME structure, I am presenting their full texts here as we find them in the Christian Science textbook on page 115.

THE SCIENTIFIC TRANSLATION OF IMMORTAL MIND (level 1)

GOD: Divine Principle, Life, Truth, Love, Soul, Spirit, Mind. *(divine synonyms)*
MAN: God's spiritual idea, individual, perfect, eternal. *(divine image)*
IDEA: An image in Mind; the immediate object of understanding. - Webster. *(divine reflection)*

THE SCIENTIFIC TRANSLATION OF MORTAL MIND (level 2, 3, 4)

Third Degree (Note: the third step in progression): **Understanding.** *(reality – 2nd level)*
SPIRITUAL. Wisdom, purity, spiritual understanding, spiritual power, love, health, holiness.
(Note: "In the third degree mortal mind disappears, and man as God's image appears.")

Second Degree: Evil beliefs disappearing. *(transitional qualities- 3rd level)*
MORAL. Humanity, honesty, affection, compassion, hope, faith, meekness, temperance.

First Degree: Depravity. *(unreality 4th level)*
PHYSICAL. Evil beliefs, passions and appetites, fear, depraved will, self-justification, pride, envy, deceit, hatred, revenge, sin, sickness, disease, death.

With the rows now fully defined, the stage is set to focus onto the columns, to define their functions. Mary Baker Eddy has provided numerous definitions that define the characteristics of the 4 columns of the ACME structure. The definitions are summarized below.

Mary Baker Eddy's definition for the 4 columns

The Word	Christ	Christianity	Divine Science
incorporeal	divine	supreme	infinite
Pison (river)	Gihon (river)	Hiddekel (river)	Euphrates (river)
Northward	Eastward	Southward	Westward
HEAL THE SICK	RAISE THE DEAD	CLEANSE THE LEPERS	CAST OUT DEAMONS

The 1st **type** is presented in the chapter The Apocalypse.
The 2nd **type** is contained in the definition for God in Recapitulation
The 3rd **type** is contained in the Glossary as a definition of the 4 biblical rivers in Genesis 2.
The 4th **type** is presented in the chapter The Apocalypse, text shown below.
Northward, its gates open to the North Star, the Word, the polar magnet of Revelation; **eastward**, to the star seen by the Wisemen of the Orient, who followed it to the manger of Jesus; **southward**, to the genial tropics, with the Southern Cross in the skies, - the Cross of Calvary, which binds human society into solemn union; **westward**, to the grand realization of the Golden Shore of Love and the Peaceful Sea of Harmony.
An additional four phrases, in this case, that also apply to the 4 columns are the 4 elements that Mary Baker Eddy has inscribed in her seal surrounding the cross and crown symbol.

Relating the various works by Mary Baker Eddy to the ACME structure.

It should be noted that Mary Baker Eddy created all of her progressive development-type structures in 16 parts each, which thereby cover the entire acme structure. The uniformity easily relates them to one-another, and to the ACME structure as their foundation.

4	8	12	16
3	7	11	15
2	6	10	14
1	5	9	13

The sequence of applications shown here applies to the textbook chapters, the Lord's Prayer stanzas, the *Christ and Christmas* verses and illustrations, and to the Church Manual segments. This means that Mary Baker Eddy's illustrated poem, Christ and Christmas, which is presented in this book needs to be seen in the above sequencing.

The Platform of Christian Science, however, that Mary Baker Eddy created in 32 parts, poses a slight challenge. It is a large structure that covers all 16 ACME elements in groups of 2 parts of the platform per element. Of course it does so in the manner of a platform, beginning at the highest position.

7, 8	5, 6	3, 4	1, 2
15, 16	13, 14	11,12	9, 10
23, 24	21, 22	19, 20	17, 18
31, 32	29,39	27, 28	25, 26

It needs to be noted here that the complexity increases further, when one adds to this context the 16-part **counter-platform** that one finds in the Glossary Definition for the term, **Mortal Mind**, as shown below. The details for it all are presented later in the book.

7, 8 4	5, 6 3	3, 4 2	1, 2 1
15, 16 8	13, 14 7	11,12 6	9, 10 5
23, 24 12	21, 22 11	19, 20 10	17, 18 9
31, 32 16	29,39 15	27, 28 14	25, 26 13

It needs to be noted further that the platform that we find in the textbook chapter *Recapitulation*, which comprises the entire chapter, is made up of only 24 parts. Because the chapter Recapitulation is focused on healing (it is Mary Baker Eddy's class book when she taught the healing practice), the Recapitulation platform applies only to the lower 3 rows.

Mary Baker Eddy opens each part of the Recapitulation platform with a critical question, like "What is God?" The fact that such questions are asked, precludes that the answer is known. The convention places the entire Recapitulation platform onto the level below the level for the TRANSLATION OF IMMORTAL MIND, because on the top level Truth is known. No question needs to be asked there. The result places the healing platforms below the top row, into the 3 lower rows.

7, 8	5, 6	3, 4	1, 2
15, 16	13, 14	11,12	9, 10
23, 24	21, 22	19, 20	17, 18

The platform of the **Bible Lesson Topics** is closely related to the Recapitulation platform. However, it is slightly larger. It is made up of 26 parts, according to the 26 weeks in half a year. Mary Baker Eddy solved the problem by placing 2 questions into the chapter Recapitulation, which address two separate topics. The dual questions are #20 and #22.

Mary Baker Eddy further created a counter platform for the Bible Lesson Topics with the 26-part, first definition, for the name, "Adam," in the Glossary. These 26 counter parts are specific elements of the Adam lie that the lesson topics overturn.

A dialogue unfolds here. The questions presented in Recapitulation are asked by mortal mind, the small-minded thinking that is typically a void in terms of knowing the truth. When questions are asked, mortal mind is 'speaking.' The answer to the question that silences mortal mind, is given by the Christ, scientifically expressed in Christian Science. In the resulting dialog, the associated platform of the Bible Lesson Topics adds to the dialog that is advancing the healing, both individually and collectively. The dialog leads out of the mortal mind dream, bringing vistas of the divine scene into the scene of the mental void. The mortal-mind void is further highlighted by the specific void that flows from the name, Adam. The dynamics are such that the Recapitulation question that is asked at the beginning, becomes obsolete by the resulting 'knowing' the truth

Where do we go from here?

In the biblical book, *Revelation*, there was war in heaven against the 'woman', against humanity, and the woman prevailed. This vision was followed up by John's vision of the new Jerusalem, the foursquare city from God becoming established on Earth. The destiny of humanity is thereby, as John saw it, clothed with the sun. This means that progress is the law of God. It comes with an imperative that we cannot escape. Progress is the nature of our being, both symbolically and in practice. We are builders of worlds, and the creators of ever-greater expression of the divine future.

Like the Universe is not finite, which expresses God, so God is not finite in any respect, but is without limits. Its fullness is the substance that is reflected in man, the image of God. It nourishes the world by the divine process of Love being reflected in love at every point in human existence. Divine Love born out as love in loving is an element of the acme of all that is good, and of Christian Science that inspires us to participate in the process of divine reflection. With it, each step opens up new opportunities in the divine participation. Thus, expressing Love cannot be anything narrow, small, and confined, but is inherently universal in its light, in order to reflect the divine.

Universal love is not something that belongs to the future. It belongs to the now and builds the future. Seeds are we; ideas creating new ideas and ever-grander worlds.

I wrote a poem once on the subject, with the title: Harvest is Seedtime. I then discovered years later that its progression matches the dimension and characteristic of my series of novels, *The Lodging for the Rose,* which explores the Principle of Universal Love. Some time later still, I discovered that the series itself matches the characteristic of the lower 3 rows of the ACME structure, as do all platforms focused on healing. None of which was intended. Healing, evidently, has a natural principle standing behind it that is universally knowable.

For a long time I had thought that my 12-part series of novels, because of its focus on the Principle of Universal Love, should pertain to the upper three rows. It is evident now that this perception was mistaken for several reasons. 1. It would render the divine, too small. 2. The fact that the series of novels was written to explore the question, what is universal love reflected at the grassroots level, indicates that the divine is far from being known and understood, but needs to be explored and understood, and then be acknowledged in living. 3. At the very best the explorations reflect a daring, exciting as this may be, to reach beyond the limits of small-mined concepts - a daring that is powered by a widening sense of love.

However, my placing the novels onto the last three rows, does not mean that they must have the character of a platform. The novels are progressive in nature all the way through. This dictates their placing on the ACME structure as a development structure. Once a principle is understood, it dictates the form of its expression. This leaves the field wide open for infinite development. Mary Baker Eddy did not provide an example for a 12-part development type structure. The progression in healing is evidently as wide as the sea shore in individual terms, unfolding from the congregation in society towards divine Truth.

Book 3	Book 6	Book 9	Book 12
Book 2	Book 5	Book 8	Book 11
Book 1	Book 4	Book 7	Book 10

I have presented the novels near the end of the book, under the summary title, *The Lodging for the Rose*.

About Christ and Christmas

An illustrated poem by Mary Baker Eddy.

The poem, *Christ and Christmas,* and its illustrations with associated elements were authored by Mary Baker Eddy as a single book, evidently for reasons of its great significance. The book was first published in 1893. The seal shown here is Mary Baker Eddy's seal that identified her publications at the time, from 1881 until 1909.

The Illustrations for the poem were produced by Mary Baker Eddy and James F. Gilman; artists. When the book, *Christ and Christmas*, was first published, it met with such confusion in the field that it was withdrawn for a time. After it was republished, Mary Baker Eddy is said to have predicted that the work would likely disappear from the shelves of the Reading Rooms before its significance would become recognized. While I cannot confirm this statement, the present 'fact' seems to bear it out to a large degree. The book was evidently for future ages – not futuristic in time, but in scientific development in society.

The book becomes immensely profound when it is seen as a work of metaphoric reference for the biblical city foursquare as a scientific platform for the development of Christian Science itself. Evidence suggests that the foursquare platform was an established central element of her work right from the beginning, leading up to 1881 when she published the constituent elements of it in her book, A Key to the Scriptures, and also published the visual metaphor for it that illustrates its complexity, which became the cross and crown seal shown above. The seal incorporates all the critical features that relate the glossary-type presentation in A Key to the Scriptures to the foursquare structure, or *the city of our God*, as she referred to it later. The crown of the 1881 seal was incorporated into the last illustration in *Christ and Christmas*.

The editorial work of bringing the poem, *Christ and Christmas*, together with the numerous related elements that Mary Baker Eddy has created, which the poem thereby illustrates, has been performed by Rolf A. F. Witzsche, with editorial references added.

The page-layout in the first part of the book, presenting Christ and Christmas, differs slightly from the original, earlier versions, such as the 1916 Edition. In the edition presented here, a number of interspersed blank pages have been omitted, whereby the verses of the poem are presented opposite the illustrations, and the repetition of the last two lines is presented below the illustration itself instead of on a separate page.

The chosen arrangement for this book, of the poem Christ and Christmas as a part of a work of references, makes the poem more easily readable in PDF form, and its linkage to the Biblical city foursquare more easily recognizable.

For the same reason is the final verse in Christ and Christmas, attributed to Christ Jesus, being brought forward in the book into the position of the last verse. This editorial action places the verse, which Mary Baker Eddy has attributed to Christ Jesus in front the list of verses authored by Mary Baker Eddy and their glossary. By this editing, the final verse of Christ Jesus referring to the MORNING STAR, is also linked with the illustration of the great star - the 'Morning Star' - that Mary Baker Eddy presented on the book's cover, placed as a illustration opposite to the verse referring to it.

Christ and Christmas

by Mary Baker Eddy

(a development-type structure)

4	8	12	16
3	7	11	15
2	6	10	14
1	5	9	13

4000

Fast circling on, from zone to zone, —
Bright, blest, afar, —
O'er the grim night of chaos shone
One lone, brave star.

STAR OF BETHLEHEM

O'ER THE GRIM NIGHT OF CHAOS SHONE
ONE LONE, BRAVE STAR.

3000

    In tender mercy, Spirit sped
      A loyal ray
    To rouse the living, wake the dead,
      And point the Way —

2000

    The Christ-idea, God anoints —
      Of Truth and Life;
    The Way in Science He appoints,
      That stills all strife.

CHRIST HEALING

THE WAY IN SCIENCE HE APPOINTS,
THAT STILLS ALL STRIFE.

1ooo

What the Beloved knew and taught,
  Science repeats,
Through understanding, dearly sought,
  With fierce heart-beats;

SEEKING AND FINDING

THROUGH UNDERSTANDING, DEARLY SOUGHT,
WITH FIERCE HEART-BEATS;

o4oo

Thus Christ, eternal and divine,
  To celebrate
As Truth demands, — this living Vine
  Ye demonstrate.

o3oo

For heaven's Christus, earthly Eves,
  By Adam bid,
Make merriment on Christmas eves,
  O'er babe and crib.

CHRISTMAS EVE

MAKE MERRIMENT ON CHRISTMAS EVES,
O'ER BABE AND CRIB.

o2oo

　　　Yet wherefore signalize the birth
　　　　Of him ne'er born?
　　　What can rehearse the glorious worth
　　　　Of his high morn?

CHRISTMAS MORN

WHAT CAN REHEARSE THE GLORIOUS WORTH
OF HIS HIGH MORN?

o1oo

       Christ was not crucified — that doom
         Was Jesus' part;
       For Sharon's rose must bud and bloom
         In human heart. *

oo4o

       Forever present, bounteous, free,
         Christ comes in gloom;
       And aye, with grace towards you and me,
         For health makes room.

* "God was manifest in the flesh." — St. Paul.

CHRISTIAN SCIENCE HEALING

AND AYE, WITH GRACE TOWARDS YOU AND ME,
FOR HEALTH MAKES ROOM.

oo3o

Thus olden faith's pale star now blends
  In seven-hued white!
Life, without birth and without end,
  Emitting light!

I THANK THEE, O FATHER, LORD OF HEAVEN AND EARTH, BECAUSE THOU HAST HID THESE THINGS FROM THE WISE AND PRUDENT, AND HAST REVEALED THEM UNTO BABES. – *Christ Jesus*

LIFE, WITHOUT BIRTH AND WITHOUT END,
EMITTING LIGHT!

oo2o

The Way, the Truth, the Life — His word —
  Are here, and now
Christ's silent healing, heaven heard,
  Crowns the pale brow.

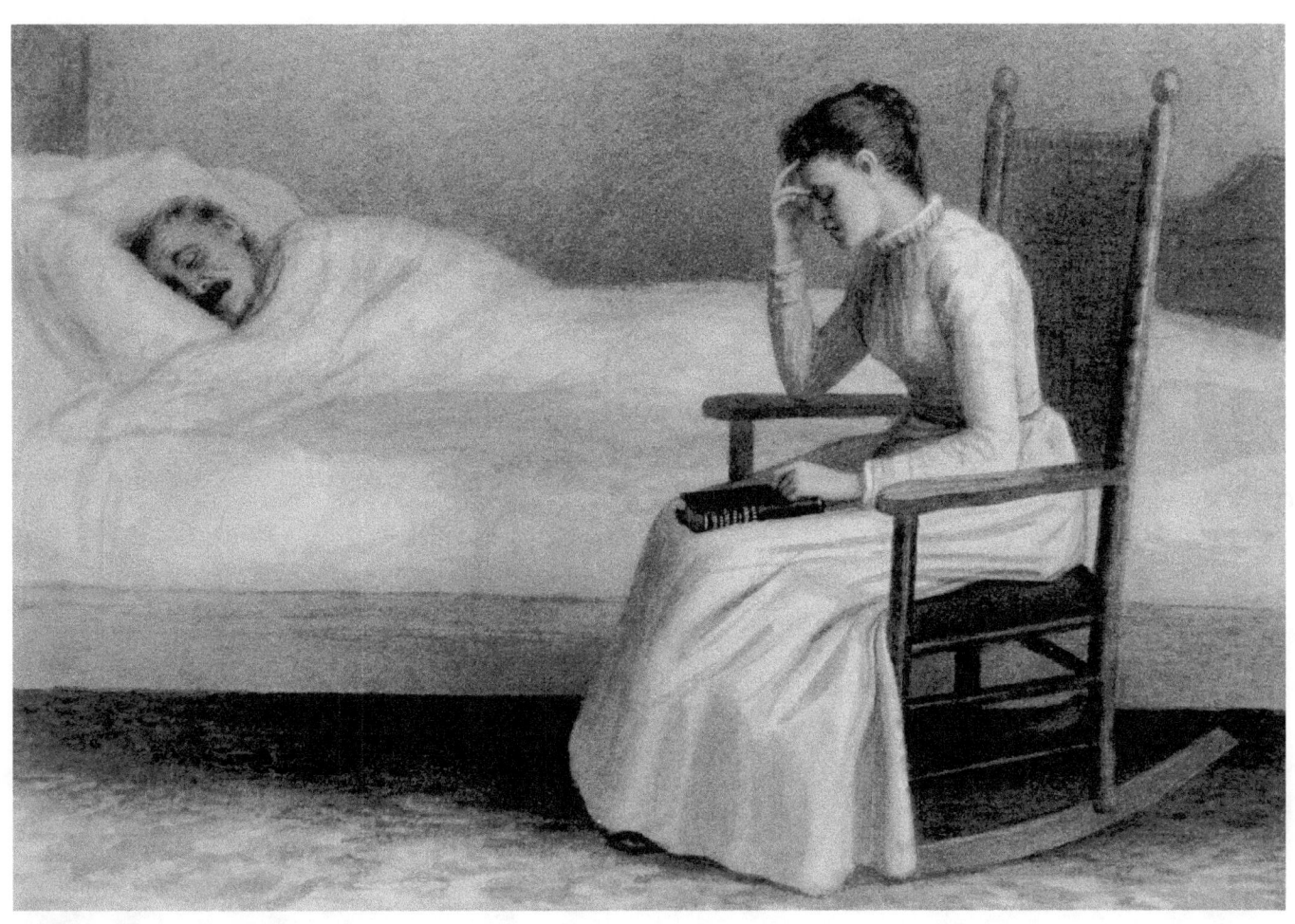

TREATING THE SICK

CHRIST'S SILENT HEALING, HEAVEN HEARD,
CROWNS THE PALE BROW.

oo1o

>For Christian Science brings to view
>   The great I Am, —
>Omniscient power, — gleaming through
>   Mind, mother, man.

ooo4

>As in blest Palestina's hour,
>   So in our age,
>'T is the same hand unfolds His power,
>   And writes the page.

CHRISTIAN UNITY

'T IS THE SAME HAND UNFOLDS HIS POWER,
AND WRITES THE PAGE.

ooo3

     To-day, as oft, away from sin
       Christ summons thee!
     Truth pleads to-night: Just take Me in!
       No mass for Me!

TRUTH VERSUS ERROR

TRUTH PLEADS TO-NIGHT: JUST TAKE ME IN!
NO MASS FOR ME!

ooo2

No blight, no broken wing, no moan,
  Truth's fane can dim;
Eternal swells Christ's music-tone,
  In heaven's hymn.

THE WAY

ETERNAL SWELLS CHRIST'S MUSIC-TONE,
IN HEAVEN'S HYMN.

ooo1

And he that overcometh, and keepeth my
works unto the end, to him will I give
power over the nations:
And I will give him the MORNING STAR.
— *Christ Jesus*.

(Part of Front Cover Image)

AND I WILL GIVE HIM THE MORNING STAR.
— *CHRIST JESUS*.

Christ and Christmas

1   1. Fast circling on, from zone to zone, —
         Bright, blest, afar, —
3   O'er the grim night of chaos shone
         One lone, brave star.

     2. In tender mercy, Spirit sped
6   A loyal ray
         To rouse the living, wake the dead,
         And point the Way —

9   3. The Christ-idea, God anoints —
         Of Truth and Life;
         The Way in Science He appoints,
12   That stills all strife.

     4. What the Beloved knew and taught,
         Science repeats,
15   Through understanding, dearly sought,
         With fierce heart-beats;

     5. Thus Christ, eternal and divine,
18   To celebrate
         As Truth demands, — this living Vine
         Ye demonstrate.

21   6. For heaven's Christus, earthly Eves,
         By Adam bid,
         Make merriment on Christmas eves,
24   O'er babe and crib.

     7. Yet wherefore signalize the birth
         Of him ne'er born?
27   What can rehearse the glorious worth
         Of his high morn?

     8. Christ was not crucified — that doom
30   Was Jesus' part;
         For Sharon's rose must bud and bloom
         In human heart. *

33   9. Forever present, bounteous, free,
         Christ comes in gloom;
         And aye, with grace towards you and
me,
36   For health makes room.

     10. Thus olden faith's pale star now
blends
         In seven-hued white!
39   Life, without birth and without end,
         Emitting light!

     11. The Way, the Truth, the Life — His
word —
42   Are here, and now
         Christ's silent healing, heaven heard,
         Crowns the pale brow.

45   12. For Christian Science brings to view
         The great I Am, —
         Omniscient power, — gleaming through
48   Mind, mother, man.

     13. As in blest Palestina's hour,
         So in our age,
51   'T is the same hand unfolds His power,
         And writes the page.

     14. To-day, as oft, away from sin
54   Christ summons thee!
         Truth pleads to-night: Just take Me in!
         No mass for Me!

57   15. No blight, no broken wing, no moan,
         Truth's fane can dim;
         Eternal swells Christ's music-tone,
60   In heaven's hymn.

     * "God was manifest in the flesh."
     — St. Paul.

GLOSSARY

These Scriptural texts are the basis of the sentiments in the verses, whereto their number corresponds.

VERSE

1. I am the root and the offspring of David, and the bright and morning star. — Christ Jesus.

2. Verily, verily, I say unto you, The hour is coming, and now is, when the dead shall hear the voice of the Son of God: and they that hear shall live. — Christ Jesus.

3. The people that walked in darkness have seen a great light: they that dwell in the land of the shadow of death, upon them hath the light shined. — Isaiah.

4. But seek ye first the kingdom of God, and His righteousness; and all these things shall be added unto you. — Christ Jesus.

5. The tabret, and pipe, and wine, are in their feasts: but they regard not the work of the Lord, neither consider the operation of His hands. — Isaiah.

6. Man that is born of a woman is of few days, and full of trouble. — Job.

7. Before Abraham was, I am. — Christ Jesus.

8. If Christ be in you, the body is dead because of sin; but the Spirit [God-likeness] is life because of righteousness. — St. Paul.

9. But such as I have give I thee: In the name of Jesus Christ of Nazareth rise up and walk. — St. Peter.

10. Without father, without mother, without descent, having neither beginning of days, nor end of life; but made like unto the Son of God. — St. Paul.

11. Heal the sick. — Christ Jesus.

12. For whosoever shall do the will of my Father which is in heaven, the same is my brother, and sister, and mother. — Christ Jesus.

13. And there shall be one fold, and one shepherd. — Christ Jesus.

14. Behold, I stand at the door, and knock: if any man hear my voice, and open the door, I will come in to him, and will sup with him, and he with me. — Christ Jesus.

15. And whosoever liveth and believeth in me shall never die. — Christ Jesus.

Christian Science and The City of our God

A City Foursquare

A city foursquare has 16 regions within it. Mary Baker Eddy created all of her major works as constituents of this 'city', including her poem *Christ and Christmas* that she created in 16 verses. In this manner, her illustrated poem stand as a 16-part visual metaphor for the 16 elements of the city foursquare that she referred to as the *city of our God*, and the "acme" of Christian Science. By this connection *Christ and Christmas* relates to all of her works.

Mary Baker Eddy defined the 16 elements both as horizontal rows, in terms of 4 levels of thinking, and in terms of 4 vertical columns that represent upwards development.

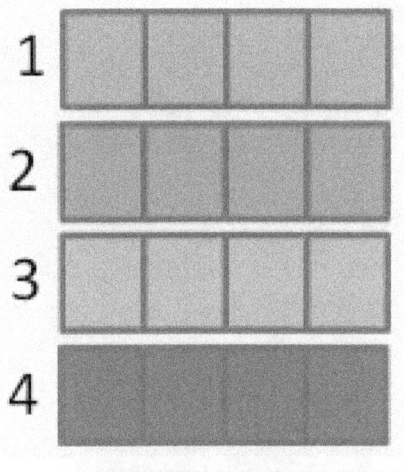

The rows are defined as:

1 **The Word** of Life, Truth, and Love.

2 The Christ, the spiritual idea of God.

3 Christianity, which is the outcome of the divine Principle of the Christ-idea in Christian history.

4 Christian Science, which to-day and forever interprets this great example and the great Exemplar.

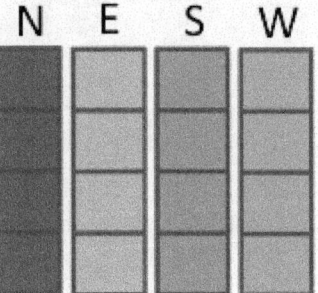

The columns are defined in terms of "sides" as:

"**The Word, Christ, Christianity**, and (now) **divine Science**". Mary Baker Eddy also refers to them as "gates" that have a geographic orientations according to the sequence of the daily cycle of the Sun, beginning northward:

Northward, its gates open to the North Star, the Word, the polar magnet of Revelation;

Eastward, to the star seen by the Wisemen of the Orient (Christ), who followed it to the manger of Jesus;

Southward, to the genial tropics, with the Southern Cross in the skies, - the Cross of Calvary, which binds human society (Christianity) into solemn union;

Westward, to the grand realization of the Golden Shore of Love and the Peaceful Sea of Harmony (divine Science).

Referring to the ACME of Christian Science

From the textbook chapter: The Apocalypse

Revelation xxi. 9: -

And there came unto me one of the seven angels which had the seven vials full of the seven last plagues, and talked with me, saying, Come hither, I will show thee the bride, the Lamb's wife.

This ministry of Truth, this message from divine Love, carried John away in spirit. It exalted him till he became conscious of the spiritual facts of being and the "New Jerusalem, coming down from God, out of heaven," - the spiritual outpouring of bliss and glory, which he describes as the city which "lieth foursquare." The beauty of this text is, that the sum total of human misery, represented by the seven angelic vials full of seven plagues, has full compensation in the law of Love. Note this, - that the very message, or swift-winged thought, which poured forth hatred and torment, brought also the experience which at last lifted the seer to behold the great city, the four equal sides of which were heaven-bestowed and heaven-bestowing.

Think of this, dear reader, for it will lift the sackcloth from your eyes, and you will behold the soft-winged dove descending upon you. The very circumstance, which your suffering sense deems wrathful and afflictive, Love can make an angel entertained unawares. Then thought gently whispers: "Come hither! Arise from your false consciousness into the true sense of Love, and behold the Lamb's wife, - Love wedded to its own spiritual idea." Then cometh the marriage feast, for this revelation will destroy forever the physical plagues imposed by material sense.

The city foursquare 575

This sacred city, described in the Apocalypse (xxi. 16) as one that "lieth foursquare" and cometh "down from God, out of heaven," represents the light and glory of divine Science. The builder and maker of this New Jerusalem is God, as we read in the book of Hebrews; and it is "a city which hath foundations." " The description is metaphoric. Spiritual teaching must always be by symbols. Did not Jesus illustrate the truths he taught by the mustard-seed and the prodigal? Taken in its allegorical sense, the description of the city as foursquare has a profound meaning. The four sides of our city are the Word, Christ, Christianity, and divine Science; "and the gates of it shall not be shut at all by day: for there shall be no night there." This city is wholly spiritual, as its four sides indicate.

The royally divine gates 575

As the Psalmist saith, "Beautiful for situation, the joy of the whole earth, is mount Zion, on the sides of the north, the city of the great King." It is indeed a city of the Spirit, fair, royal, and square. Northward, its gates open to the North Star, the Word, the polar magnet of Revelation; eastward, to the star seen by the Wisemen of the Orient, who followed it to the manger of Jesus; southward, to the genial tropics, with the Southern Cross in the skies, - the Cross of Calvary, which binds human society into solemn union; westward, to the grand realization of the Golden Shore of Love and the Peaceful Sea of Harmony.

Revelation's pure zenith 576

This heavenly city, lighted by the Sun of Righteousness, - this New Jerusalem, this infinite All, which to us seems hidden in the mist of remoteness, - reached St. John's vision while yet he tabernacled with mortals.

The shrine celestial 576

In Revelation xxi. 22, further describing this holy city, the beloved Disciple writes: -

And I saw no temple therein: for the Lord God Almighty and the Lamb are the temple of it.

There was no temple, - that is, no material structure in which to worship God, for He must be worshipped in spirit and in love. The word temple also means body. The Revelator was familiar with Jesus' use of this word, as when Jesus spoke of his material body as the temple to be temporarily rebuilt (John ii. 21). What further indication need we of the real man's incorporeality than this, that John saw heaven and earth with "no temple [body] therein"? This kingdom of God "is within you," - is within reach of man's consciousness here, and the spiritual idea reveals it. In divine Science, man possesses this recognition of harmony consciously in proportion to his understanding of God.

Divine sense of Deity 576

The term Lord, as used in our version of the Old Testament, is often synonymous with Jehovah, and expresses the Jewish concept, not yet elevated to deific apprehension through spiritual transfiguration. Yet the word gradually approaches a higher meaning. This human sense of Deity yields to the divine sense, even as the material sense of personality yields to the incorporeal sense of God and man as the infinite Principle and infinite idea, - as one Father with His universal family, held in the gospel of Love. The Lamb's wife presents the unity of male and female as no longer two wedded individuals, but as two individual natures in one; and this compounded spiritual individuality reflects God as Father-Mother, not as a corporeal being. In this divinely united spiritual consciousness, there is no impediment to eternal bliss, - to the perfectibility of God's creation.

The city of our God 577

This spiritual, holy habitation has no boundary nor limit, but its four cardinal points are: first, the Word of Life, Truth, and Love; second, the Christ, the spiritual idea of God; third, Christianity, which is the outcome of the divine Principle of the Christ-idea in Christian history; fourth, Christian Science, which to-day and forever interprets this great example and the great Exemplar. This city of our God has no need of sun or satellite, for Love is the light of it, and

divine Mind is its own interpreter. All who are saved must walk in this light. Mighty potentates and dynasties will lay down their honors within the heavenly city. Its gates open towards light and glory both within and without, for all is good, and nothing can enter that city, which "defileth, . . . or maketh a lie."

The writer's present feeble sense of Christian Science closes with St. John's Revelation as recorded by the great apostle, for his vision is the acme of this Science as the Bible reveals it.

- *Psalm XXIII* - 577

In the following Psalm one word shows, though faintly, the light which Christian Science throws on the Scriptures by substituting for the corporeal sense, the incorporeal or spiritual sense of Deity: - PSALM XXIII

[Divine Love] is my shepherd; I shall not want. Love] maketh me to lie down in green pastures: Love] leadeth me beside the still waters. [Love] restoreth my soul [spiritual sense]: [Love] leadeth me in the paths of righteousness for His name's sake.

Yea, though I walk through the valley of the shadow of death, I will fear no evil: for [Love] is with me; [Love's] rod and [Love's] staff they comfort me. [Love] prepareth a table before me in the presence of mine enemies: [Love] anointeth my head with oil; my cup runneth over.

Surely goodness and mercy shall follow me all the days of my life; and I will dwell in the house [the consciousness] of [Love] for ever.

Written by Mary Baker Eddy

The City Foursquare as a scientific development structure

Development, by its nature, is upwards oriented; from primitive or erroneous concepts towards truthful scientific concepts. This means that the development in the columns flows from the lowest element in a column to the highest element. Since the columns themselves are progressive from left to right, the overall progression proceeds column by column, as illustrated in Christ and Christmas.

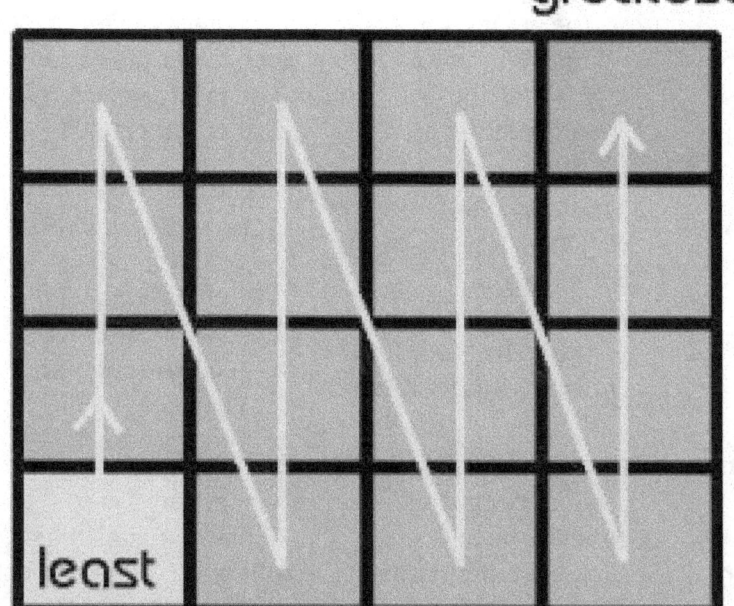

greatest

least

Developmental Structure

A number of structures have been created by Mary Baker Eddy as progressive structures that would apply to the 'city foursquare' (the science-city of our God) in the development orientation shown here, where the development in one column aids the progression in the next column

Mary Baker Eddy's works that apply in this manner are:

The 16 illustrated verses of the poem Christ and Christmas
The 16 chapters of the Christian Science textbook.
The 16 stanzas of the Lord's Prayer
The 16 segments of the Church Manual

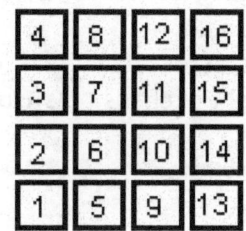

The 'city foursquare' brings all of the above together as expressions of a underlying science, reflecting the universal nature of truth. By their relationship that each of the above have to the 16 elements of the 'city common common foursquare', the individual elements of the individual structures all relate to one another across the various structures. It opens up new dimensions for exploration. This also means that the various metaphors that Mary Baker Eddy presented in Christ and Christmas are directly relevant and illustrative of the related structures, such as textbook chapters, the Lord's Prater stanzas, and also to the segments of the Church Manual.

Since the columns unfold in 4 progressive stages. the individual elements in the columns are also horizontally interrelated. The scientific interrelationship that the horizontal linking creates, brings to light a much enriched meaning for the individual elements in each of the four structures.

It is amazing what comes to light when one looks at Christ and Christmas in this manner, and the related structures, the textbook chapters, the Lord's Prayer stanzas, and the segments of the Church Manual.

The verses in Christ and Christmas

4. What the Beloved knew and taught, Science repeats, Through understanding, dearly sought, With fierce heart-beats;	8. Christ was not crucified-- that doom Was Jesus' part; For Sharon's rose must bud and bloom In human heart. [*]	12. For Christian Science brings to view The great I Am,-- Omniscient power,-- gleaming through Mind, mother, man.	And he that overcometh, and keepeth my works unto the end, to him will I give power over the nations: And I will give him the MORNING STAR. --Christ Jesus.
3. The Christ-idea, God anoints-- Of Truth and Life; The Way in Science He appoints, That stills all strife.	7. Yet wherefore signalize the birth Of him ne'er born? What can rehearse the glorious worth Of his high morn?	11. The Way, the Truth, the Life--His word-- Are here, and now Christ's silent healing, heaven heard, Crowns the pale brow.	15. No blight, no broken wing, no moan, Truth's fane can dim; Eternal swells Christ's music-tone, In heaven's hymn.
2. In tender mercy, Spirit sped A loyal ray To rouse the living, wake the dead, And point the Way--	6. For heaven's Christus, earthly Eves, By Adam bid, Make merriment on Christmas eves, O'er babe and crib.	10. Thus olden faith's pale star now blends In seven-hued white! Life, without birth and without end, Emitting light!	14. To-day, as oft, away from sin Christ summons thee! Truth pleads to-night: Just take Me in! No mass for Me!
1. Fast circling on, from zone to zone,-- Bright, blest, afar,-- O'er the grim night of chaos shone One lone, brave star.	5. Thus Christ, eternal and divine, To celebrate As Truth demands,--this living Vine Ye demonstrate.	9. Forever present, bounteous, free, Christ comes in gloom; And aye, with grace towards you and me, For health makes room.	13. As in blest Palestina's hour, So in our age, 'T is the same hand unfolds His power, And writes the page.

[*] "God was manifest in the flesh." -- *St. Paul*.

The illustrations in Christ and Christmas

4 - Seeking and Finding

8 - Christian Science Healing - part 1

12 - Christian Unity part 1

16 - The 'Morning Star'

3 - Christ Healing part 2

7 - Christmas Morn

11 - Treating the Sick

15 - The Way

2 - Christ Healing part 1

6 - Christmas Eve part 2

10 - I thank thee oh Father...

14 - Truth *Versus* Error

1 - Star of Bethlehem

5 - Christmas Eve part 1

9 - Christian Science Healing – part 2

13 - Christian Unity part 2

Interrelationship of Elements

The Christian Science textbook, Science and Health with Key to the Scriptures, is evidently not an end in itself, but

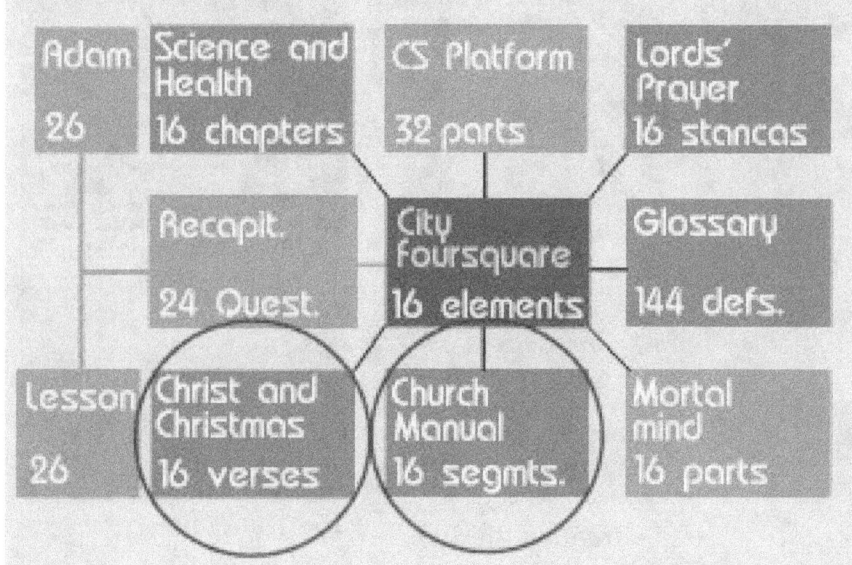

has been designed as a constituent part of the 'city foursquare' that John the Revelator described as, coming down from God out of heaven, which Mary Baker Eddy referred to as *"the city of our God."*

On this note, Mary Baker Eddy has designed all her major works to reflect the scientific characteristic of the 4-square structure. The textbook, the Lord's Prayer, and the segments of the Church Manual, do this in the form of developmental structures. They do this not in a subservient fashion, but because the 4-square structure is a powerful scientific foundation for advanced exploration, which evidently had been already recognized in Jesus' time, as John the Revelator indicates and may have been utilized by both John and Jesus in their own explorations.

By their global interrelationship to the 4-square foundation, all individual elements also become interrelated with each other by the nature of the scientific process. Christian Science is not a religion to be taken on faith, but is a science that develops understanding and advanced concepts and capabilities. Mary Baker Eddy evidently expected increasing progress along this line.

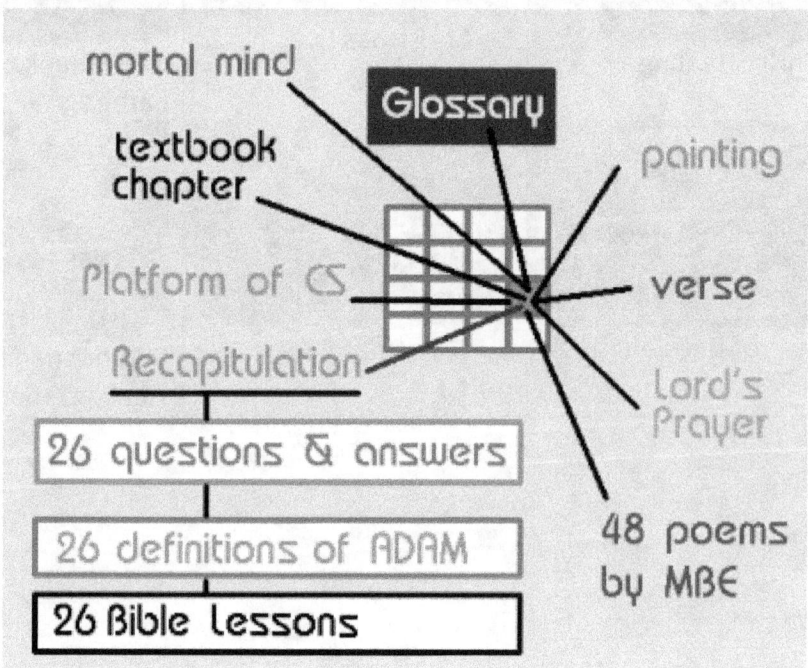

The Chapters of the Christian Science Textbook

4 - Christian Science versus Spiritualism	8 - Footsteps of Truth	12 - Christian Science Practice	16 - The Apocalypse
3 - Marriage	7 - Physiology	11 - Some Objections Answered	15 - Genesis
2 - Atonement and Eucharist	6 - Science Theology Medicine	10 - Science of Being	14 - Recapitulation
1 - Prayer	5 - Animal Magnetism Unmasked	9 - Creation	13 - Teaching Christian Science

The Lord's Prayer

Adorable One.	Enable us to know, - as in heaven, so on earth, - God is omnipotent, supreme.	And Love is reflected in love;	For God is infinite, all-power, all Life, Truth, Love, over all, and All.
Hallowed be Thy name.	Thy will be done in earth, as it is in heaven.	And forgive us our debts, as we forgive our debtors.	For Thine is the kingdom, and the power, and the glory, forever.
Our Father-Mother God, all-harmonious,	Thy kingdom is come; Thou art ever-present.	Give us grace for to-day; feed the famished affections;	And God leadeth us not into temptation, but delivereth us from sin, disease, and death.
Our Father which art in heaven,	Thy kingdom come.	Give us this day our daily bread;	And lead us not into temptation, but deliver us from evil;

The Church Manual, as a model for the advance of civilization

The Church Manual, the Manual of the Mother Church, The First Church of Christ Scientist in Boston, Massachusetts, by Mary Baker Eddy, is far more than a collection of by-Laws and rules for the government of the church. It stands as a platform for the government of a society by constitutional principles which become powerfully productive when the principles are scientifically understood, and thereby become an impelling and also enabling, power.

With this factor in mind, the Church Manual is designed as a development-type structure, rather than as a platform, and needs to be so applied to the 'city foursquare' accordingly, as a constituent of it; as a part of the ACME structure. The 16 segments of the Manual by their nature, more than by the specifics in the by-laws, have a universal significance for the scientific recognition of the dynamics of civilization. By these dynamics the elements are both vertically and horizontally interrelated. The scope of the dynamics is so large that books can be written about them, and likely will in future times, which are merely indicated here by the topics of the Manual's 16 unique segments.

The full text of the Manual (except the Appendix) has been added to the Appendix of this book for reference purposes. It is not the purpose of this book to explain the significance of the various segments, but to make their structural interrelationships, both horizontally and vertically, more 'visible'.

It is noteworthy, however, that the Manual hints at the existence of three levels of education in Christian Science, a "primary" level for education in the field; a "normal" level for the education of the teachers in the field carried out in the Board of Education that awards the degree CSB (a bachelor type degree); and a higher degree CSD that may be taken at the Massachusetts Metaphysical College that functions under Mary Baker Eddy as President in perpetuity (symbolically), without teachers attached, and under a symbolic President, which awards no certificates. However, Mary Baker Eddy notes in the marginal heading of the sample application form in the Appendix of the Manual that a person who has "taken a degree at the Massachusetts Metaphysical College" is qualified to countersign a membership application. She seems to suggest that the highest degree cannot be certified by anyone, for who would be the judge at the leading edge? Let the works therefore stand as testament that a person has taken a degree (symbolically) with Mary Baker Eddy at the College. And she seems to say further to the applicant for membership in the Mother Church that it is every person's own critical duty to be keenly aware of who 'underwrites' one's significant steps in life. (See segment, Board of Education)

The Manual of the Mother Church - Segments

4. MEETINGS	8. THE MOTHER CHURCH AND BRANCH CHURCHES	12. BOARD OF EDUCATION	16: CHURCH MANUAL
3. DISCIPLINE	7. RELATION AND DUTIES OF MEMBERS TO PASTOR EMERITUS	11. TEACHING CHRISTIAN SCIENCE	15: CHURCH-BUILDING
2. CHURCH MEMBERSHIP	6. READING ROOMS	10. THE CHRISTIAN SCIENCE PUBLISHING SOCIETY	14: COMMITTEE ON PUBLICATION
1. CHURCH OFFICERS	5. CHURCH SERVICES	9. GUARDIANSHIP OF CHURCH FUNDS	13. BOARD OF LECTURESHIP

Platform Structures

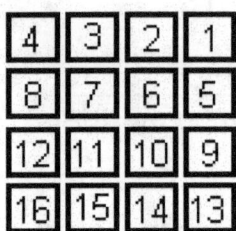

Platform structures are different in nature from developmental structures. Platform structures begin with the highest element of recognition and then present everything below it that leads up to the highest recognition.

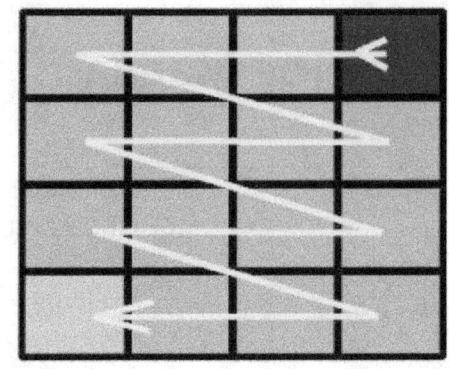

Some platforms are extremely complex. The Christian Science Platform, for example, is made up of 32 parts. Evidence suggests that Mary Baker Eddy has organized the Platform into groups of two parts per element.

Mary Baker Eddy has also created a 16-part counter-platform, which she presents in the Glossary of the textbook as her scientific definition for the small-minded sense opposing reality, termed "mortal mind." The Platform of Christian Science overturns the "mortal mind" notions, or imposition, item by item.

The Counter-Platform – Mortal Mind

4 - a suppositional material sense, alias the belief that sensation is in matter, which is sensationless;	3 - error creating other errors;	2 - mythology;	1 - Nothing claiming to be something, for Mind is immortal;
8 - the belief that man is the offspring of mortals;	7 - the belief that life has a beginning and therefore an end;	6 - the opposite of Spirit, and therefore the opposite of God, or good;	5 - a belief that life, substance, and intelligence are in and of matter;
12 - material senses;	11 - the subjective states of error;	10 - idolatry;	9 - the belief that there can be more than one creator;
16 - death.	15 - sickness;	14 - sin;	13 - that which neither exists in Science nor can be recognized by the spiritual sense;

The Platform of Christian Science (subheadings)

7- Divine trinity	5- Allness of Spirit	3- Evil obsolete	1- The deific supremacy
8- Father-Mother	6- The universal cause	4- Life the creator	2- The deific definitions

15- The Son's duality	13- The divine Principle and idea	11- Christ Jesus	9- The Son of God
16- Eternity of the Christ	14- Spiritual oneness	12- Messiah or Christ	10- Holy Ghost or Comforter

23- Indivisibility of the infinite	21- The divine Ego	19- Soul and Spirit one	17- Infinite Spirit
24- God the parent Mind	22- The real manhood	20- The one divine Mind	18- The only substance

31- Evil not produced by God	29- Adam not ideal man	27- True idea of man	25- Man reflects the perfect God
32- Basis of health and immortality	30- Divine pardon	28- Truth demonstrated	26- Purity the path to perfection

For the complete platform statements, see the Appendix

Recapitulation, Adam, and the Lesson Topics

We have a triple platform combination here, with a complex interrelationship between the three platforms.

The main platform is Mary Baker Eddy's class book that she utilized for the teaching of her science of metaphysical healing. The class book became incorporated into the textbook of Christian Science as the chapter, *Recapitulation*. The entire chapter is a platform by design, as it begins with the highest element. However, it is a unique platform in that it covers only 3 rows of the 4-square structure, which it is a part of. The chapter contains 24 parts, which renders it a 2-part structure for each of the 12 elements of the three rows that it covers.

The three-row short form is required for the chapter to correlate with the Bible Lesson Topics, because when a calendar year is divided into two halves, 26 weekly Bible lessons are required to cover this time frame. The 24-part Recapitulation platform has been created in a manner that can accommodate this requirement. The chapter thereby provides an interface between the Bible lessons and the 4-square structure. In order for the interface to work, Mary Baker Eddy has structured two of the 24 Recapitulation questions in a manner that accommodates two elements. The Recapitulation elements 20 and 22 are of this type.

Mary Baker Eddy also created a 26-part counter-platform that stands in opposition to the 26-part Bible Lesson Platform. The counter platform consists of the first part of the Glossary definition for the name, Adam. This is a hugely long definition, presented as a 26-part sentence that has little meaning by itself. However, it becomes significant as a counter-platform against the Bible Lesson Topics. The counter-platform presents specific aspects of the Adam-lie-against-man that the Bible Lessons are designed to purge from consciousness. The Recapitulation platform serves as a resource for the Bible Lessons, and for overcoming the Adam dream, that is to heal our thinking of it at every point. For this reason, the three platforms function as a single unit, as I see it.

.

As a healing structure, the 24 parts of the triple platform cannot actually be larger than the space of the TRANSLATION OF MORTAL MIND, which covers the three lower rows. The healing process is evidently not a physical process, but a process of freeing oneself and humanity from the trap of a MORTAL MIND mentality, which is not legitimate. The triple platform is designed to break the mortal-mind bondage to small-minded thinking, and raise it up to evermore reflect the translation of IMMORTAL MIND, individually and universally. This is my present perception of the dynamics involved, according to what I see laid out for us by Mary Baker Eddy. The Lesson Topics raise humanity out of the mortal-mind dream from all its levels as represented by ADAM, to the level of God that is the only level legitimate for man's identity.

Please note, the Appendix of this book contains the complete chapter *Recapitulation* with the Bible Lesson topics added, together with the applicable part of the definition for the name, Adam.

On the following page, the three elements, Recapitulation, Adam, and the Lesson Topics, are combined, element by element, and related to the 4-square structure, coincident as I see it, with Mary Baker Eddy's TRANSLATION OF MORTAL MIND where healing is required to eliminate the mortal concepts.

Since the Adam concept plays a role here, I see the second part of the Glossary definition for the name Adam as 4 specific denials of God that pertain to the four columns. "The name Adam represents the false supposition (1) that Life is not eternal, but has beginning and end; (2) that the infinite enters the finite, that intelligence passes into non-intelligence, and that Soul dwells in material sense;(3) that immortal Mind results in matter, and matter in mortal mind; (4) that the one God and creator entered what He created, and then disappeared in the atheism of matter

7. - What is substance? **Adam**: A belief in intelligent matter, finiteness, and mortality. **7 - SOUL**	**5**. - What are the demands of the Science of Soul? **Adam**: The opposite of good,--of God and His creation. **5 - LOVE**	**3**. - Is there more than one God or Principle? **Adam**: The belief in "original sin," sickness, and death **3 - LIFE**	**1**. - What is God? **Adam**: Error **1 - GOD**
8. - What is Life? **Adam**: "Dust to dust" **8 - MIND**	**6**. - What is the scientific statement of being? **Adam**: A curse **6 - SPIRIT**	**4**. - What are spirits and souls? **Adam**: Evil **4 - TRUTH**	**2**. - Are these terms synonymous? **Adam**: A falsity **2 - SACRAMENT**
15. - What are body and Soul? **Adam**: A co-called finite mind, producing other minds, thus making "gods many and lords many" **15 - ARE SIN, DISEASE, AND DEATH REAL?**	**13**. - Is there no sin? **Adam**: The opposite of Spirit and His creations **13 - REALITY**	**11**. - Are doctrines and creeds a benefit to man? **Adam**: The first god of mythology **11 - SUBSTANCE**	**9**. - What is intelligence? **Adam**: Red sandstone **9 - CHRIST JESUS**
16. - Does brain think, and do nerves feel, and is there intelligence in matter? **Adam**: A product of nothing as the mimicry of something **16 - DOCTRINE OF ATONEMENT**	**14**. - What is man? **Adam**: That which is not the image and likeness of good, but a material belief, opposed to the one Mind, or Spirit **14 - UNREALITY**	**12**. - What is error? **Adam**: Not God's man, who represents the one God and is His own image and likeness **12 - MATTER**	**10**. - What is Mind? **Adam**: Nothingness **10 - MAN**
23. - How can I progress most rapidly in the understanding of Christian Science? **Adam**: Immortality's opposite, mortality **25 - IS THE UNIVERSE, INCLUDING MAN, EVOLVED BY ATOMIC FORCE?**	**21**. - Do the five corporeal senses constitute man? **Adam**: Life's counterfit, which ultimates in death **22 - ANCIENT AND MODERN NECROMANCY, ALIAS MESMERISM AND HYPNOTISM, DENOUNCED**	**19**. - Is materiality the concomitant of spirituality, and is material sense a necessary preliminary to the understanding and expression of Spirit? **Adam**: An inverted image of Spirit **19 - ADAM AND FALLEN MAN**	**17**. - Is it important to understand these explanations in order to heal the sick? **Adam**: An unreality as opposed to the great reality of spiritual existence and creation **17 - PROBATION AFTER DEATH**
24. - Have Christian Scientists any religious creed? **Adam**: That of which wisdom saith, "Thou shalt surely die." **26 - CHRISTIAN SCIENCE**	**22a**. - Will you explain sickness... **Adam**: The opposite of Love, called hate **23 - GOD THE ONLY CAUSE AND CREATOR** **22b** - and show how it is to be healed? **Adam**: The userper of Spirit's creation, called self-creative matter **24 - GOD THE PRESERVER OF MAN**	**20a**. - You speak of belief. **Adam**: The image and likeness of what God has not created, namely, matter, sin, sickness, and death **20 - MORTALS AND IMMORTALS** **20b**. - Who or what is it that believes? **Adam**: The opposer of Truth, termed error **21 - SOUL AND BODY**	**18**. - Does Christian Science, or metaphysical healing, include medication, material hygiene, mesmerism, hypnotism, theosophy, or spiritualism? **Adam**: A so-called man, whose origin, substance, and mind are found to be the antipode of God, or Spirit **18 - EVERLASTING PUNISHMENT**

The foundation that Christian Science is built on

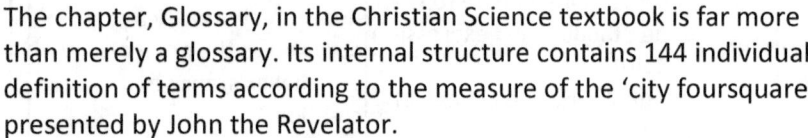

frontal few = 9 jewels
on 5 stems

The chapter, Glossary, in the Christian Science textbook is far more than merely a glossary. Its internal structure contains 144 individual definition of terms according to the measure of the 'city foursquare' presented by John the Revelator.

global view
contains
16 jewels

Mary Baker Eddy's application of the 144 concepts to the 16 elements of the 'city foursquare' (9x16=144) is illustrated in metaphor in Christ and Christmas in the painting Truth versus Error. The angel that we see knocking at the door of humanity with the scroll of Christian Science in hand, stands in the middle of two concentric squares that are woven into the carpet. The metaphor yields a 9-part concentric structure for each of the 16 elements surrounding a central concept.

The 9x16 dimension is also indicated in the design of the crown that presents 9 jewels in frontal view, and 16 in total. The structure of the crown was first published in 1881, at a very early stage of the development of Christian Science, at which point the Glossary was essentially complete, published at the time under the title, *Key to the Scriptures*. Mary Baker Eddy's seal that incorporates the 9x16 crown, which she applied to her works from 1881 to 1908, illustrates the nature of the foundation that Christian Science is built on.

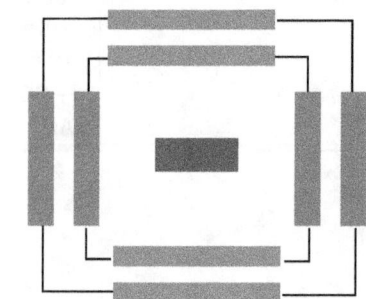

The Glossary Structure of 144 parts

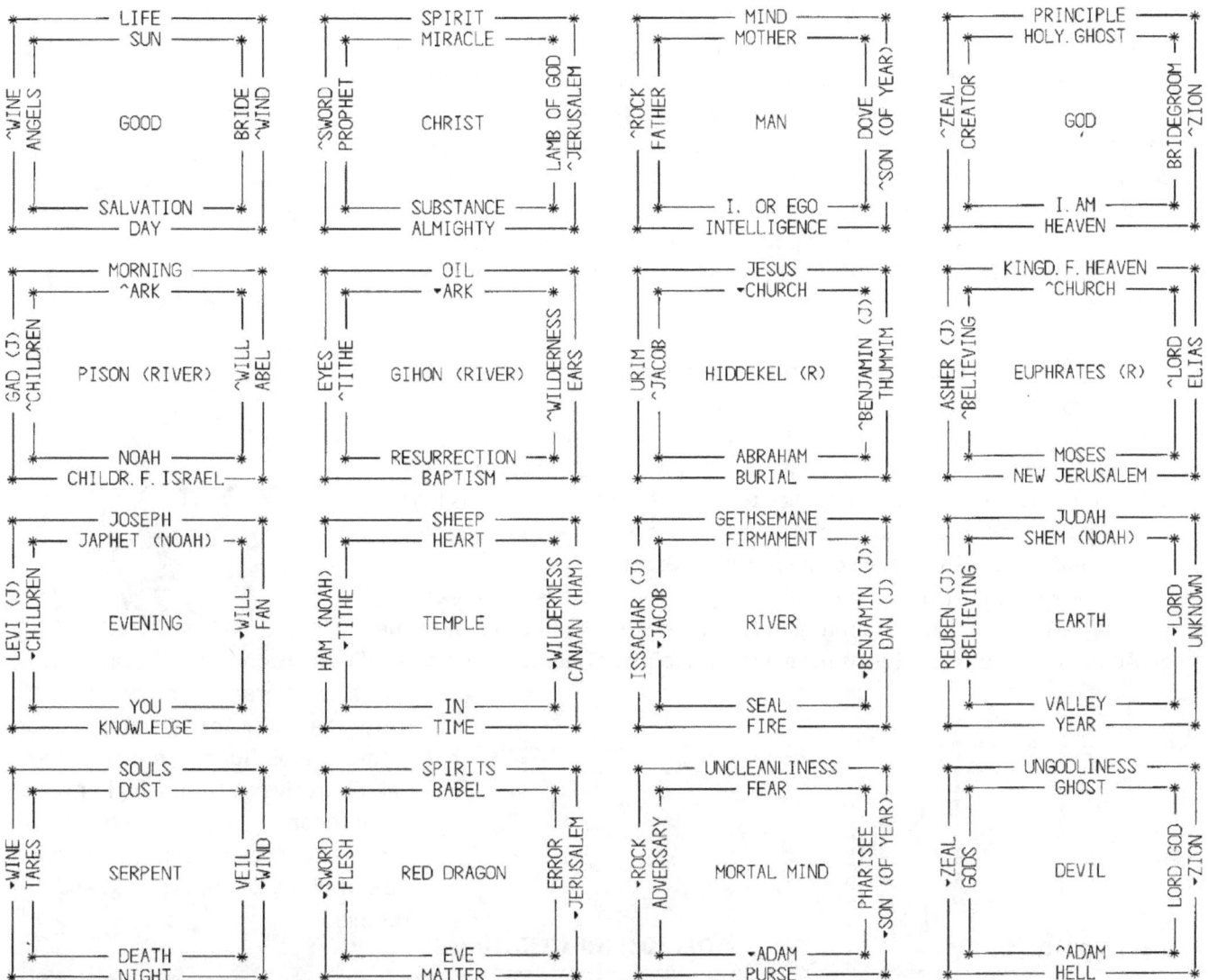

The above is an example of the type of application of the glossary definitions to the 16-elements of the 'city foursquare' that Mary Baker Eddy had likely been working with, according to the evidence that she has presented. She referred to the 4-square structure, referencing John's vision (or perception) of It, as *the city of our God*, and the "acme" of Christian Science. (S&H 577:30)

In her final edition of the Christian Science textbook, Science and Health with Key to the Scriptures, the Glossary that contains the above definitions, begins immediately following the paragraphs under the sub-heading, *The city of our God*. In earlier editions (the 214th edition, in 1901) the sub-heading was, *Compass and light*, which she later changed to, *The city of our God*. Under the same subheading she presents an example with the 23rd Psalm, of "the light which Christian Science throws on the Scriptures by substituting for the corporeal sense, the incorporeal or spiritual sense of Deity." The above-illustrated example serves as a research foundation for individual, scientific, spiritual development, which, by its nature, has no final limit. The placement of the 144 definitions depends on one's own, individual progressing understanding of the science involved. Mary Baker Eddy did not provide a final product, which is not possible as science is an open-ended landscape of discoveries without final limits.

Five types of dual definitions

The chapter "Glossary" in the Christian Science textbook is not a simple collection of definitions of terms. It is far-more in the nature of an enormously challenging scientific platform. For example, it lists fewer than 144 defined terms, of which a number have been given a dual definition. A careful examination reveals that Mary Baker Eddy has presented five different types of dual definitions that all have their own significant concept to convey. This feature evidently was important to Mary Baker Eddy right from the beginning on, as the factor of 5 is prominently included into the 1881 crown of her seal and was emphasised further in 1908 when she modernized her seal, as if she was saying, "pay attention to this factor, this is big."

The most common type is that of the vertical duality where a term has both a spiritual and material significance, such as for, Zion. Eight of these are included in the Glossary. In some cases, the vertical duality is less stark so that it spans only two levels, as in the case of Benjamin. The duality of a term can also have an additional reference added, that applies to both of the dual terms, as in the case of Son, where the reference "son of a year" gives a special meaning to both the spiritual and material sense of sonship. (MIS.180:26)

When the duality is horizontal in meaning, it gives a term different meanings, but at the same level, which pertain to different contexts at that level, as in the case of, Adam. In some cases, terms have a clear duality defined which is scientifically inseparable, because the duality is inherent in the nature of the concept and cannot be separated, as in the case of, Burial. Another type of definition gives a term a higher or lower meaning, depending on the context in which it is seen, as in the case of River, Earth, Evening, and Temple.

The following page summarizes the various definitions Mary Baker Eddy has applied to the individual rows and columns.

Mary Baker Eddy presents us with terms in the glossary that are not arbitrarily chosen, but dissolve into 4 groups that precisely match the characteristics of the 4 levels of the 4-square structure. Within these groups we find various types of duality presented that are each of unique significance. When the work is done, the needed 144 definitions are recognized with none left over and none too few.

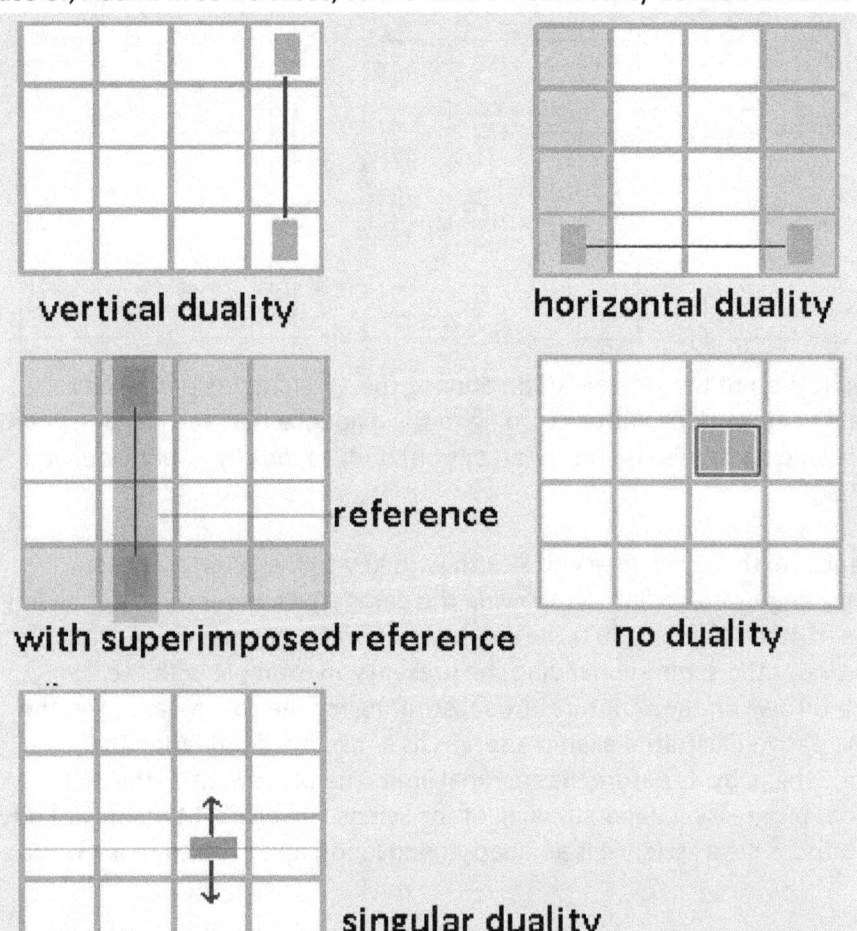

vertical duality

horizontal duality

with superimposed reference
reference

no duality

singular duality

THE VERTICAL DUALITY

Two groups of 8 terms each, with vertical duality, do exist in the Glossary. They are distinguished by their type of contrast, opposite or relative, respectively.

The **opposite-duality** group of terms contains the terms: **Jerusalem, Rock, Son, Sword, Wine, Wind, Zeal, Zion.** (See Zion as an example:) - "Zion. Spiritual foundation and superstructure; inspiration; spiritual strength. Emptiness; unfaithfulness; desolation." The duality is presented in separate sentences.

The **less-contrasting vertical-duality** group contains the terms: **Believing, Benjamin, Children, Jacob, Lord, Tithe, Wilderness, and Will.** - Note: For two terms in this group, for the terms "Benjamin" and "Children," the duality is presented in separate paragraphs, which is rare, suggesting the existence of a special interrelationship.

THE VERTICAL DUALITY WITH A SUPERIMPOSED REFERENCE.

This is a special case applied to the vertical term, **Son**, only. This special definition has the concept, "Son of a year" superimposed on both aspects of the duality (Son. The Son of God, the Messiah or Christ. The son of man, the offspring of the flesh. "Son of a year.") Mary Baker Eddy explains it this way. She defines the concept, "Son of a year", saying: "In the Hebrew text, the word 'son' is defined variously; a month is called the son of a year. This term, as applied to man, is used in both a material and a spiritual sense." (Miscellaneous Writings p.180)

THE DIRECTIONAL DUALITY

The terms that have a vertical duality indicated in a single sentence are terms that point into opposite directions, depending on their context. These are: **Earth, Evening, River, and Temple.**

THE HORIZONTAL DUALITY

The terms for which a distinct horizontal duality is indicated, are the terms: **Adam, Ark, and Church.** In these cases, the duality is separated into separate paragraphs.

THE INDIVISIBLE DUALITY

Another important dual definition presents a duality that is required to define a concept, which therefor cannot be separated, or else the concept would be lost. Mary Baker Eddy presents us 3 examples of this type in the Glossary in the definitions for: **Burial, Devil, and Euphrates.**

UNDEFINED TERMS (not found in the Glossary)

The above terms are all defined in the Glossary. However, there exists a group of three terms, **Soul, Truth, and Love**, synonyms for God, which one would expect to see defined, but which are omitted from the Glossary.

Soul, Truth, and Love

John the Revelator, in describing the city foursquare on Revelation 21, presented also the factor 3 as highly important (verse 13), together with the factors 4, 12 and 144.

13 On the east three gates;
on the north three gates;
on the south three gates;
and on the west three gates.

In her description of the city foursquare in the chapter The Apocalypse, Mary Baker Eddy describes the 'city' only in factors of 4. She didn't omit the other factors. The factor 12 became the basis for the Recapitulation Platform that is made up of 12 groups of 2 parts each. She gives the factor 12 a 'big' significant thereby. Nor did she ignore the factor 144. She built the entire Glossary structure on it. The result is' gigantic' in its significance. But what about the factor 3? John seems to suggest that it has a global significance. Mary Baker Eddy responded by giving the factor 3 the most 'majestic' significance of them all, that is deep-reaching and all-pervading. This factor is highlighted by an act of evidently intentional omission.

In the Glossary of her textbook on Christian Science, Mary Baker Eddy has defined the term, God, (in part), with the following sequence of terms.

God – Principle; Mind; **Soul**; Spirit; Life; **Truth; Love**; all substance; intelligence.

Of these, three terms, three of the synonyms for God (Soul, Truth, and Love) are not specifically defined in the Glossary. Why would these significant terms be omitted? Are they not significant enough?

The reason for the omission evidently is that these terms are of special significance in the overall design of the Glossary structure in the sequence in which the terms are given, whereby they become extremely significant, as of they were the heart of everything.

Just look at what happens when the synonyms are applied as a platform to the ACME structure, in the sequence given. The concept "Principle" becomes associated with the most significant element, with the rest following towards the left in the manner of a platform representation. This means, the concept "Mind" is next, to head up the 3rd column. Then, suddenly, a break occurs. The concept of "Soul" is not defined in Glossary, and can therefore not be placed in any position. Does this make sense? No, it doesn't. But it makes perfect sense when one recognizes the concept as a separator between the right and left halves of the ACME structure, which attributes to each of the two haves a unique significance.

With the concept of "Soul" thereby rendered with a global significance, the next two concepts, "Spirit" and "Life" are defined again in the Glossary and can therefore be placed in the top position of the second and first column respectively.

But what about the concepts "**Truth**" and "**Love**" that are next in the sequence, but are not defined in the Glossary? It becomes evident that they two have a 'global' significance to identify the nature of the two upper rows, respectively. The concept, "**Truth**", thereby is characteristic for everything that pertains to the TRANSLATION OF IMMORTAL MIND. Every part of it is **Truth**. God is **Truth**. God is immediately relevant to the nature and existence of man. God is not a part of a remote mysterium. This is immensely significant, isn't it? **Truth** is that big!

The final synonym for God in the sequence given in the Glossary is the concept, "**Love**." It, too, is immensely significant in that it defines the operating impetus in the TRANSLATION OF MORTAL MIND that the dynamics in the lower three rows facilitate. If **Love**, reflected in love in society, in an active manner, would not happen at all, civilization would not exist and everything would be chaos. In fact, nothing would exist. **Love** is that big!

In this context, on the second row, **Love** defines **the Christ**; "the spiritual idea of God," as Mary Baker Eddy has referred to the Christ. (S&H 577:15). Mary Baker Eddy does not even mention the term Christ in either the TRANSLATION OF IMMORTAL MIND and in the TRANSLATION OF MORTAL MIND. The reason may be that the dimension of the Christ, its function in the context of humanity, cannot be fully recognized without the nature of the Christ becoming recognized in **Divine Love being reflected in love**, and this connection being fully put onto the table. Still, the dynamics is described by Mary Baker Eddy in the TRANSLATION OF MORTAL MIND, because in the degree in which Love is not actively reflected in love, humanity **deprives** itself of a vital element of its existence and of civilization, whereby **depravity** becomes rampant and civilization cannot stand.

Mary Baker Eddy defines the Christ in the Glossary as: "The divine manifestation of God, which comes to the flesh to destroy incarnate error." This is the function of Love reflected in love, isn't it?

A sense of completeness.

		Truth	Life	Spirit	Soul	Mind	Principle
Substance (Lower case after 1907)							
Intelligence		Love	Pison	Gihon		Hiddekel	Euphrates
			evening	temple		river	earth
			serpent?	red dragon?		mortal mind?	hell?

With the three omitted terms that Mary Baker Eddy did not define in the Glossary, but presented with a higher significance instead, her rendering of St John's vision of the city foursquare is full completed. With this she could say that what St. John had laid onto the table as a foundation to build on, is the "acme" of "Christian Science."

The sequence of terms in the definition of God listed above has remained the same almost from the beginning (1884, maybe earlier), and so has the omission of the terms Soul, Truth, and Love in the glossary structure. The undefined terms, by their sequence, can be seen to divide the structure into two halves by Soul as the dividing line, with Truth and Love defining the upper two rows, respectively

Applying the missing terms, **Soul, Truth, and Love** to the dual structure of the scientific translation of Immortal Mind and mortal mind.

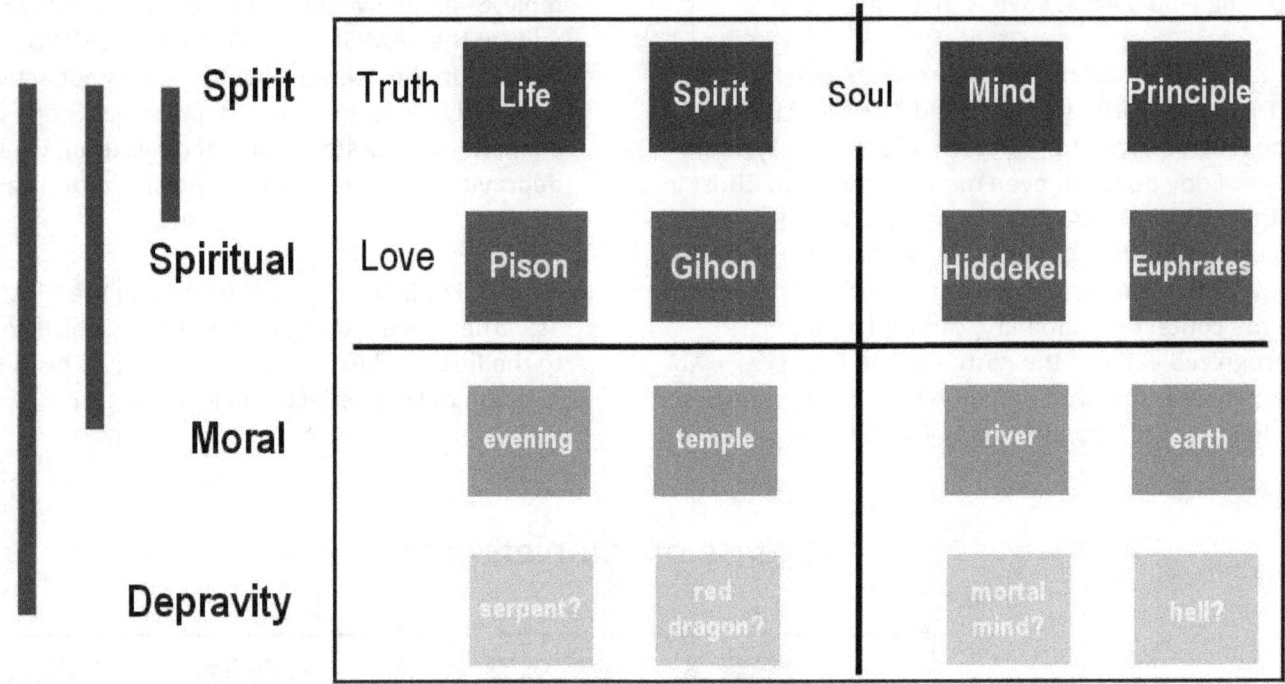

		Spirit		Truth	Life	Spirit	Soul	Mind	Principle
Spiritual		Love	Pison	Gihon		Hiddekel	Euphrates
Moral			evening	temple		river	earth
Depravity			serpent?	red dragon?		mortal mind?	hell?

The arrangement shown here, illustrates the upper row as the Translation of Immortal Mind, whereby Truth defines the domain of Spirit, and in the second row, Love, that defines the domain that is spiritual, the domain of "the Christ, the spiritual idea of God", as Mary Baker Eddy has referred to the second level.

Mary Baker Eddy defined, Man, as an element of the translation of Immortal Mind – the translation of it in life. Here man is seen as an element of Spirit. Mary Baker Eddy is reported to have once said to her secretary, "You are Spirit." He replied, "no, Mother, I am spiritual." She repeated emphatically, "You are Spirit."

The lower three rows define stages of approximation. When the real is attained, powered by the Christ expressing Love, the lower three rows, representing mortal mind, vanish as they lose their apparent reality. They become obsolete. Until this is accomplished, however, the lower three rows are significant for their development potential in the process of Christ-healing.

My perception is, that the chapter Recapitulation, of the textbook, which was Mary Baker Eddy's class book in the early stages, and its associated Bible Lesson Topics, which cover only three rows, pertain to the three rows where the dynamics of healing takes place, where Love is the top-line of the healing dynamics and reflects "the Christ, the spiritual idea of God." This means that before we are qualified to deal with the numerous parts of the Glossary structure, it appears wise to look more closely at the significance of Love as the great 'power factor' at the gate to the divine All, to the 1st row where God and man are One and all is Truth.

Temple, Ark, Church

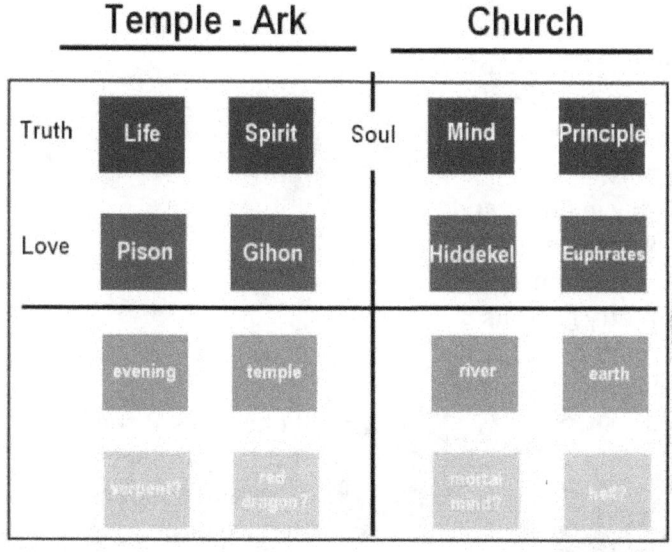

Temple - Ark   **Church**

			Soul		
Truth	Life	Spirit		Mind	Principle
Love	Pison	Gihon		Hiddekel	Euphrates
	evening	temple		river	earth
	serpent?	red dragon?		mortal mind?	hell?

.What does the division of the structure, by the term, Soul, signify?

If one looks closer, the placing of the term Soul as a concept with a larger, global, significance, separates the ACME structure into two 'mirrored' halves that have each a unique orientation. I have identified the two orientations, as an example, with the Glossary-terms Church, and Temple and Ark. Both concepts appear significant. I like to compare Temple and Church as unique orientations. However, the Glossary definitions for Ark and Church are structurally identical, and both belong to the same level (row 2, related to Love). It appears that Mary Baker Eddy presents a significant lesson here, with the term Ark (safety, the idea, or reflection of Truth...) juxtaposed with Church (the structure of Truth and Love; whatever rests upon and proceeds from divine Principle...) Are we looking at two types of safety, within and without?

I find the concept of Ark highly significant, far taller than temple, by its scientific signification (God and man coexisting and eternal...). It stands as a significant healing factor.

This dual orientation, centered on Soul, is evident in all structures that Mary Baker Eddy has created for the foursquare *city of our God*, as she has referred to it.

The dual orientation is important, in that it can serve as one of the deciding factors for determining the placement of the individual Glossary definitions within the foursquare structure. This is a 'task' that Mary Baker Eddy has placed into our court, individually. Each of the 144 concepts that comprise the Glossary structure has been designed by her to be examined with the same type of considerations that the concept of Ark invites as a potential identifier for one of the mirrored halves centered on Soul. Each of the 144 concepts in the Glossary needs to become scientifically understood, which has a significance for the whole and for the healing efficacy of Christian Science.

It appears that the derived understanding of specific aspects is designed to have a specific healing effect in each single case.

The above application of the concept of Ark in the context with Church, for example, as identifiers, invariably helps one to decide scientifically into which half a specific Glossary definition belongs. Nothing is arbitrary here. Everything is determined by the scientific significance involved. This is a challenge that may never end to inspire new insights, and also it may have been this unbounded significance, which inspired Mary Baker Eddy to refer to John's pioneering revelation of the 4-square city as the ACME of the science that she termed Christian Science.

Placing the individual Glossary definitions

There are 144 individual definitions of scientific spiritual concepts contained in the Glossary of the Christian Science textbook, according to the measure established for the foursquare 'city' by the Revelator, John. (Rev. 21:17). Since a 4-square structure is made up of only 16 elements, so that John's measure of 144 contributes 9 definitions to each of the 16 elements, further considerations are required for placing an individual definition within the 9-part structure of each of the 16 elements.

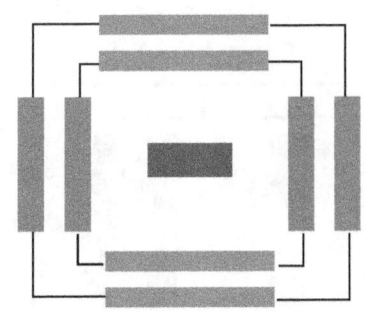

A potential arrangement has been illustrated by Mary Baker Eddy in Christ and Christmas in the scene, *Truth versus Error*. We see the angel of science, who is knocking on the door of humanity, standing at the center of two concentric squares, woven into a carpet, faintly visible. The resulting geometry gives us 8 sides to define, with the angel's position at the center, for a total of 9.

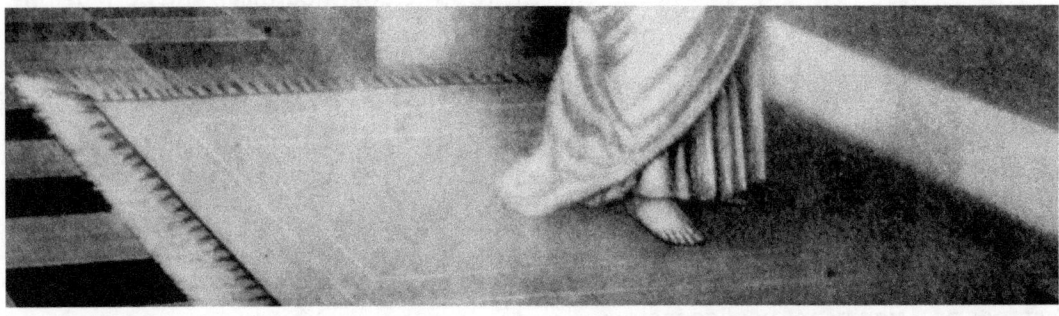

If one looks at the resulting arrangement, it becomes apparent that we have 4 horizontal positions to define that reflect in the small the nature of the four rows, and 4 vertical positions that reflect the nature of the columns, in the small, with the central position reflecting the position of the angel of science. The arrangement leaves absolutely no room for any arbitrary positioning. All is determined by the respective science involved, which is an infinite task, because science is not abstract in nature, but unfolds in life and healing and spiritual development.

Mary Baker Eddy did not provide a final product of the resulting structure that one might frame and hang on the wall. Such a final product would only be possible if science was abstract and its object was finite. But this isn't the case. Mary Baker Eddy defines, God, in her class book (Recapitulation) as "incorporeal, divine, supreme, infinite Mind, Spirit, Soul, Principle, Life, Truth, Love". Note, the terms are separated here by commas, which renders every concept listed as being infinite in nature. The 4-square structure is therefore by design, infinite in nature, forever developing, and forever developing humanity, and with it, developing its capacity for healing and for building and creating.

The only contribution that Mary Baker Eddy could make, and did make, was to outline the dynamics of the 4-square structure and provide the building blocks for it of the needed types and the needed quantities, with none missing and none left over, all nicely alphabetically sorted with none of them predetermined. The only predetermined parts that she provided are the associated structures, such as that of the textbook chapters, the Lord's Prayer stanzas, the Church Manual segments, the Platform statements, the Lesson topics, and the Recapitulation questions and answers, and so on. While the associated structures are all complete in themselves, their significance is only partially realized, and will remain so until their grounding in the 4-square structure is being recognized and understood, where their deeper significance and power is located. It may well be that the field of Christian Science will diminish to insignificance before the underlying foundation of Christian Science, in the science that Mary Baker Eddy has established for it, will become recognized and understood. Evidently, without this understanding, too much remains missing for the great potential of Christian Science to be realized for healing individuals and nations, and the world.

The Transition Line

Mary Baker Eddy defined the 3rd row as "transitional" (in her Scientific translation of mortal mind). This means that 4 Glossary definitions must exist that have a directional duality as single term that can lead in both directions. These exist. They are the definitions for Evening, Temple, River, and Earth. With these as central terms on the 3rd row, the lower 3 rows can be divided into two groups of 56 terms each, with both groups sharing the directional terms.

The dividing line appears to have been highly significant for Mary Baker Eddy, because she has surrounded the 7-poited star on the front cover of *Christ and Christmas* with 56 rays of light. The illustration in Christ and Christmas on the 3rd row, reflect this duality to some degree. But more significant may be the fact that all of the scientifically critical related structures have been provided by Mary Baker Eddy in textbook chapters located on the 3rd row, including the critical metaphor for the Glossary structure that indicated in Christ and Christmas in the same element where the chapter Recapitulation is located.

(Chapter: Atonement and Eucharist)	The scientific translation of Immortal Mind and mortal mind	The Platform of Christian Science	Recapitulation / Lessons Glossary 9x16 metaphor.

In the illustration in Christ and Christmas (Christ Healing) that coincides with the textbook chapter Atonement and Eucharist, the scene indicates that below the line the scene is essentially dead, humanity lives in a coffin. Mary Baker Eddy has labeled the scene below as "depravity". The term suggests, a state of being deprived of good and life, the typical outcome of self-deprivation. She is saying in essence to humanity, you don't have to let this happening to you. In the scientific translation of mortal mind, the Christ is pointing upwards, raising the dead out of the coffin of self-deprivation, which often the deprivation of faith in matter and sickness, instead of in Spirit, science, and health.

The Platform of Christian Science serves a similar purpose, to rouse humanity to claim its divine identity, to heal history, for the present to reflect the truth.

In Recapitulation, the voice that uplifts is the Christ's divine Science. Her placing the structures on the transition line seems to suggest that these structures all a life and death significance and should not be ignored, including the Glossary structure that presents the greatest challenge in terms of scientific honesty and awards the greatest freedom, for no one ever falls asleep on this platform. Her warning seems to be, don't fall asleep on the moral line, but reach for the top. She seems to say that there are 56 rays of light on our horizon if we stop sleeping and day the dreaming that may be summarized as mortal mind that exists only in dreams. Being awake, we consciously reflect God where reality is anchored.

The 9 birds surrounding the cross

That Mary Baker Eddy's poem, Christ and Christmas is designed to be an integral part of the 'city foursquare' is evident by a link that exists between the Glossary of the textbook and the last illustration in Christ and Christmas.

In the last illustration we see 9 dark birds surrounding the central cross. In the Glossary, Mary Baker Eddy defined 9 names of the children of Jacob. Some perch on the cross, two fly above, and one below it. In the Glossary, Mary Baker Eddy defines the names according to the motivation of Jacob's two wives for bearing the children, each vying for Jacob's love. The various types of motivations remain to the present day a part of the relationships scene between individuals and nations, and people relating to themselves as divine human beings, and to God whose image humanity bears in truth.

The nine birds in their various positions can be seen as representation of mortal mind that likewise can be seen as unfolding in stages of degrees, of a scene that requires degrees of healing. The correlation of the 9 birds has a significance for the process of lifting mortal mind out of its small-minded concepts, to accept the divinity of humanity.

The 9 birds can also represent the 9 other biblical names defined in

the Glossary, such as Elias, Moses, and Noah and his sons, etc. These other names represent the spiritual history of humanity's self-perception, which the Bible stories, to a large degree, present.

The white bird that soars above the scene, bearing the scroll of science, evidently represents the Christ, exemplified by Christ Jesus, and to some degree also by Jacob after the Penial transformation through scientific revelation.

The white bird is shown proceeding from the crown that represents the 'city foursquare' scientific structure, which appears to be critical, if not essential, to the full scope of Christ healing, and with it to the freedom of humanity. Until the healing is complete, the lack of it stands as a black cross in the foreground. Mary Baker Eddy had warned in numerous ways that Christ and Christmas would remain an enigma open to speculation. This is a natural consequence without its connection with the 'city foursquare' being recognized. This, *Christ and Christmas* illustrates in metaphor as the black cross. It is a shadow of healing not yet achieved.

The Significance of Seven

Mary Baker Eddy has consistently limited herself to listing the seven terms, Mind, Spirit, Soul, Principle, Life, Truth, and Love, as synonymous terms for God, in her definitions for the term God. Throughout her writings, she uses a number of other capitalized terms, referring to God in various contexts, such as the term Being (S&H 466:1), or Deity (446:20). Why would she limit herself to the 7 terms that she uses in her definitions for the term God? That the seven-fold nature is significant, is illustrated by the fact that all stars in Christ and Christmas are 7-points stars. Typically, symbolic stars are six-pointed stars, like the star of David, and later, Israel.

Christ and Christmas does not reveal the secret further, except on its front cover where the 7-pointed star is surrounded by 56 rays of light. Arithmetically, 56 divided by seven yields 8. Indeed, Mary Baker Eddy's Glossary definitions include 8 dual definitions with a complete opposite vertical duality, and 8 more with a lesser vertical duality. However, a greater reason may stand behind her use of the 7-fold symbolism, which appears to be coincident with the 7 days of creation, which may be seen in metaphor as the dynamic sequence of the unfolding dimension of science itself as shown below.

First day	The infinite has no beginning - it signifies "the only" God is light - Light preceeds the sun. Upon the dark void without form, shone light...	The forever demand "let there be light"	Principle	The Word	
		recognition	The great I AM	omnipotence	
Second day	The "firmament" appears that divides the waters, above and below a demarkation line, separating the scientific and spiritual from the erroneous and mythological (including matter and limitations)	Intelligence vs. darkness Science vs. error	Mind	The Christ - spiritual	
		understanding	the all-knowing	omiscience	
Third day	The separation of the waters and dry land - (thought organized by purpose) - herb yielding seed, and trees, fruit - creation is ever appearing from the nature of its inexhaustible source.	exalted thoughts - discovery of fundamental principles	Soul	Christianity - moral	
		reflection	the all-seeing	omnipresence	
Fourth day	Lights in the 'firmament' appear. A greater (divine Science) to rule the "day", and a lesser (Christian Science) to deal with the "night" of error. In divine Science God is revealed as infinite light.	subdivision of thought ordering of perception	Spirit	Christian Science - physical	
		comprehension	the all-acting	omni-action	
Fifth day	This stage brought to the scene life in the 'waters', and birds in the air - pure and 'winged' ideas, and the charge for them to multiply - to be productive and fill their domains with life.	the dynamics of life: growth and abundance	Life	Civilization	
		multiplication	the all-wise	economy	
Sixth day	At this stage the metaphor of unfolding focuses on the quality of thought - 'cattle' (stern resolve), and 'beasts' (the lion of the tribe of Judah), and the highest idea "man" (the Son of God, having dominion and the capacity to replenish the earth).	replenish the earth	Truth	Humanity	
		dominion	the all-loving	truthfulness	
Seventh day	At its fruition the unfolded idea must meet with acceptance, and an acknowledgement of its value and completeness (rest). Completeness is not a limit to infinity, it defines infinity in quality - God and man coexist in the divine, infinite calculus of Love.	rest in action	Love	Renaissance	
		acknowledgement	the eternal	sublimity	

The seven days of creation - stages of unfolding of a spiritual idea.

Synonyms and definitions for the term God from the Glossary of Science and Health.

The Greatest Scientific Statement of Our Time

From all of her writings that Mary Baker Eddy has produced, one single statement was selected that is read from the podium at the end of every Sunday Service of the Mother Church of Christ Scientist that Mary Baker Eddy has founded, and likewise became a practice at every Branch Church of Christ Scientist.

The statement is Mary Baker Eddy's answer to the question she posed in the chapter, Recapitulation:

What is the scientific statement of being?

Her answer is a 6-part statement:

1. There is no life, truth, intelligence, nor substance in **matter**.
2. All is infinite Mind and its infinite manifestation, for God is All-in-all.
3. **Spirit is** immortal Truth; **matter** is mortal error.
4. **Spirit is** the real and eternal; **matter** is the unreal and temporal.
5. **Spirit is** God, and man is His image and likeness.
6. Therefore man is not material; he is spiritual.

The statement is centered on a contest between two opposite perceptions, of either Spirit being all, or all being matter. Where do we find the evidence located?

The statement that all is Spirit has inspired many a healing. The experienced healings stand as significant evidence of Spirit being real. Can the same be said about matter? Do we have any evidence that matter is real? Countless claims have been raised that matter is real. But is it? Where is the proof?

Since too many false concepts abound, about the nature of man, let's raise the bar and ask the question, what is real, concerning the universe. Is the universe a universe of Spirit or of matter? To which of the two does the evidence point?

Mainstream science tells us that the cosmic universe is totally material. It speaks of a sun as an isolated entity that exists on its own, is powered by its own resources that it consumes, which are deemed material through and through, in the form of hydrogen gas atoms that are heated and compressed by gravity with such force that some atoms fuse together and produce energy as they do so. But is this true? Or is it but a fairy tale that is told to children to put them to sleep? But where is the

evidence? No evidence exists.

Now let us look at the universe as a universe of Spirit. That is where we find all the evidence that points to something real. According to leading-edge science in plasma physics, the universe is a universe of plasma. Plasma is theorized to consist of electric particles, termed protons and electrons, which themselves are made up of quarks that are recognized to be but moving points of energy that are organized by a vast array of harmonizing creative principles. Not a speck of matter is recognized to exist. The lack of such a recognition puts mainstream science into a quandary.

Since energy is known to have no mass, plasma should have no mass either. But plasma is known to have mass. To solve the quandary, it is theorized that a 'matter'-type particle must exist within the protons from which their mass is derived. The theorized 'matter'-particle that is deemed to give plasma its mass, is termed, the Higgs Boson, according to the name of the man who invented the theory. The term "boson" has been coined to refer to theorized particles. In an attempt to prove that the "Higgs" theorized 'matter' particle exists, and other similar enigma, a gigantic experiment has been built in Europe at the CERN laboratory at a cost of over a trillion Swiss francs. The largest part of it, the Large Hadron Collider at CERN is a circular particle accelerator 27 kilometers long, built 100 feet underground, with 4 experiment stations on its path that are gigantic in themselves, one alone weighing 7,000 tonnes, being several times larger in size than a house.

These are not small efforts. The scale of these efforts reflects a type of desperation to prove a theory that has no basis in fact, such as the theory that basic matter does exist. The CERN institute has a staff of more than 2,000, and 12,000 associates that have become drawn to the project from 600 universities and research institutions. The Large Hadron Collider is designed to smash large quantities of protons together at a closing speed of nearly double the speed of light, hoping for the small chance that a Higgs boson will appear. The method that is applied here is basically a fascist approach, rather than a method to discover the truth. The collider creates conditions that don't exist naturally in the universe; it causes the universe to squirm, so to speak. It is foolish to aim to discover from it what is normal. The Higgs that pops out in the experiment, then becomes essentially a product that is synthesized by the experiment itself.

In December 2011, two teams at the Large Hadron Collider, who were looking for the Higgs Boson, had seen results that could suggest that the Higgs Boson exists. After more than a year of further testing, the teams announced that they now think the discovered particle is truly a Higgs Boson.

Whatever was discovered turned out to be NOT a 'matter' particle, but a boson that obeys the conservation of energy law, which states that no energy is created or destroyed, but is instead transformed, such as in the case at hand, into the phenomenon of mass whereby energy takes on the quality of mass. It merely reflects what Einstein had predicted more than half a century earlier.

The bottom line is that the gigantic effort at CERN did not prove that basic matter does in fact exist. Rather it comes as close as one can get to proving conclusively that basic-matter simply does NOT exist in any shape or form, so that the proof of it would be the only case in the history of the world where the term "miracle" would actually apply.

Real miracles, which are by definition phenomena for which no principles exist, simply don't happen. This also means, inversely, that all apparent 'miracles,' such as the healing effects of the Christ in humanity, are deeply rooted natural phenomena. In the case of Christ-healing, the healing results from the recognition of underlying divine Principle reflected in man, becoming intelligently applied. The result may seem miraculous to a society locked into small-minded perceptions, but in real terms the result is as natural as the rain watering the Earth to nourish life. In Science, no real miracle has ever happened, which means that a matter particle will never be found. It would be a miracle indeed, if this would happen - a phenomenon without a cause. In Science, cause is exclusively the 'property' of Spirit. Monumental evidence exists that this is so. It is seen everywhere. The universe is Spirit manifest in countless forms. In physics, its name is plasma.

Plasma is simply energy in motion, organized by harmonizing, creative principles that form, power, maintain, and advance everything that we term collectively the physical universe. There, all is harmony; creative and progressive harmony. There, every sun is a sphere of plasma that has interstellar streams of plasma flowing into it, which combine to create atomic forms that furnish the 'canvass' for the painting of life across it. Still, there is no matter in an atom. An atom is a construct that is synthesized out of essentially, nothing. Its shape is 100 million times larger in size than the size of the plasma electrons that create its shape. The atoms in turn harmonize with one-another in a near unlimited variety of creative interaction with which the worlds are formed, such as the one that has become our home. But there is no matter in any of it. Everything is Spirit manifest in created forms produced by boundless Intelligence. We ourselves stand at the pinnacle of the creative process, with the power to reflect the creative principles of the universe in our own living and our own creativity, and our own power to express intelligence. We are literally at one with the universe, even as the universe is one and reflects the Intelligence, Spirit, and the power of Principle that created it, which we term God. By reflection we are one with God.

"Principle and its idea is one," writes Mary Baker

Eddy. Nothing is isolated in the Universe that is Spirit, through and through. Nothing stands on its own, even physically. Physically, all stars that are a sun, exist bound to one-another in vast arrays of interstellar plasma streams that pervade the galaxies, by which they exist and are powered, just as the galaxies themselves are linked in cosmic networks of intergalactic plasma streams that ultimately power them with the energies that exist latent in space and becomes localized as needed for the universal purpose. This truth reflects the leading-edge perception of the functioning of the universe, though it is far from being acknowledged in the darkened halls of science where matter is hailed as king.

Mary Baker Eddy's revolutionary "scientific statement of being," is scoffed at in the darkened hall, which are darkened by small-minded perception that has not yet been healed, where a healing is not even contemplated.

In the darkened halls, where Mary Baker Eddy's revolutionary statement is scoffed at, our sun, for example, is deemed to be a sphere of hydrogen gas, a giant sphere of matter in which matter acts upon matter and squeezes out energy in the process by which a sun is deemed to consume itself towards it eventual death. That's a lullaby song that is effective in putting children to sleep. But when one looks for evidence for the fairy tale, one finds that no such evidence exists, except in dreams. No real actual physical evidence exists for the fairy tale, because the tale, simply put, isn't true. Only evidence contrary to the fairy tale is known to exist, as it must be, since this evidence reflects what is known to be true.

Since not a speck of matter is involved in the entire creative process in the universe, from a star to a man, our future is shaped by how the actual dynamics, the dynamics of Spirit is perceived, understood, acknowledged, and responded to. This means that Mary Baker Eddy's revolutionary, "scientific statement of being" stands at the center of hugely conflicting ideologies, with equally large effects on civilization. Thus, her statement stands with the greatest-ever significance for civilization and for the very existence of mankind.

In material theory, the Sun as a materialistic solar process, as impossible as it is, renders the Sun a rock solid invariable factor for the climate on Earth. On this rock-solid, self-maintained, invariable platform, Ice Ages are deemed to be phenomena that physics itself renders essentially impossible, based on the belief in an invariable Sun. Since a large body of evidence exists that Ice Ages are real and have been the normal state of the world for up to 85% of the last half-million years, exotic theories have been invented to render the impossible, plausible, but imagined to be still thousands of years distant, so that no preparations for it are deemed necessary at the present. That's the universally held theory. However, the theory, being based on mechanistic materialistic assumptions, is build on a premise that is not rue at its very core.

The physical evidence, which reflects the characteristics of the known dynamics of plasma-physics in the universe, renders our Sun a variable star. Electric resonance characteristics are an inherent feature in plasma dynamics. Evidence for it exists both in the small and in the large. Measurements of this evidence, made on the Earth, and also in space by a satellite orbiting the Sun, tell us that a transition is on progress towards a phase-shift in the solar dynamics that will likely occur in the 2050s or slightly before, by which the surface temperature of the Sun will rabidly diminish from the present 5,800 degrees (K) to roughly 4,000 degrees, which results in 70% less radiated solar energy received on the Earth. The resulting phenomenon matches the historic pattern observed in ice core explorations in Greenland and Antarctica. The measurements all together present evidence of a living, resonating, dynamic universe, rather than a universe that is dead, according to materialistic perception, and a Sun that is self-consuming matter and is dying.

Naturally, the dynamic nature of the universe requires us to be dynamic in our living as human beings, instead of allowing ourselves to fall asleep and die in the easy chair of rigid materialistic conventions. Divine Science tells us that reality is Spirit, and Spirit is dynamic, "forever developing itself from its boundless basis." Nothing is static there. Life is not static. Material perception alone is static, because there, nothing is real.

The coming Ice Age in the 2050s is so severely feared in the materialistic circus of small-minded perception that it is hidden from sight by all passible means. Back in the early 1970s the scientific community has been concerned about the steps that would need to be taken to protect human living in the radically transforming world in an ice age environment. The concern was put to sleep in 1974 with the (anti-human) manmade-global-warming doctrine, where nothing is real, where materialistic dreams are conjured up, without real evidence in support of them.

In the resulting 'sleep' state, no imperative is allowed that would inspire humanity to become dynamic in its living to match the cosmic dynamics. In fact, the sleep state has been maintained for so long already that any call for arousal is seen as an imposition. The first call for arousal was uttered already 150 years ago by Mary Baker Eddy when she composed her "scientific statement of being" and later caused the statement to be read from the podium at every Sunday service in the church that she founded on its basis. The power of this statement remains yet to be recognized. The day of this happening will be when humanity frees itself from the present materialistic trap and becomes creatively dynamic with a power that reflects its omnipotence. The responding to the Ice Age Challenge that stands before mankind requires a massively creative dynamism, on a scale that enables the building of 6,000 new cities for a million people each to facilitate the relocation of society out of the regions that the Ice Age phase shift will render uninhabitable, into the tropics, and much of it afloat onto the seas. This, of course, includes the relocation of all associated agriculture and industries, and the building of vast new transportation, energy, and water-supply systems. The requirements on this scale point to a level of dynamics that is not achievable on a materialistic platform that is rich in small-minded thinking, but is achievable on the basis of the creative power of infinite Spirit reflected in the creative actions of a humanity that is becoming spiritually alive.

The needed construction projects don't pose a huge physical challenge. The technologies already exist, and so do the materials for them and the energy resources. With the utilization of large-scale, automated industrial processes and high temperature nuclear power, the needed vast infrastructures can all be produced within a couple of decades if the appropriate efforts are made. All this can be completed in 30 years. But will we do it?

This is not a question of, can we do it? This can be done with relative ease, because creative efforts, even on this large scale, are not materialistic in nature. Matter creates nothing. The human spirit is the creator, reflecting divine Spirit that is not measurable in term of scale, but in term of infinity.

How we respond to the Ice Age Challenge before us, is essentially a question of ideology, not physics. In one ideology all is matter, self-consuming, dying, and humanity stands impotent and indifferent, even to its own critical needs. In this ideology nothing is created to meet future needs, whereby humanity commits itself to death in 30 years. In the other ideology, nothing is matter, and the creative spirit of humanity is unfettered, boundless, and powerfully productive on the wings of grand ideas. In this divine ideology that is the native home of man, the continued living of humanity is assured, and this in a rich renaissance environment. Thus the question of Spirit or matter is enormously huge in consequences as well as in opportunities depending on the choice that is made. The choice is between universal death in small-minded dreaming, and the freedom of reality by taking hold of it.

There the universe is recognized as a plasma universe that is intensely creative, instead of self-consuming. The conflict echoes almost line by line the scientific statement of being that Mary Baker Eddy pioneered almost 150 years ago when the concept of plasma hadn't been discovered yet. The modern discoveries, of course, have given the statement vastly greater significance. That's the difference in significance built into the statement by Mary Baker Eddy uttered 150 years ago, which remains the most significance statement today for this very reason. It will remain that until the underlying issue is resolved in the course of spiritual healing, or humanity dies out to a large degree by the consequences of its failing to achieve this healing.

When the "scientific statement of being" was first

presented, its primary importance was its role in scientific metaphysical healing. Now it plays a critical role in assuring the continued existence of 99% of humanity as a whole.

Should humanity choose the spiritual universe as reality, the impending glaciation period would cause no problems in its living. A sane humanity, grounded in the Truth, would simply relocate itself into the tropics where the remaining sunshine promises to be sufficient for its living to continue, and this with ever-greater forms of prosperity. As I said, humanity would create for itself a network of new cities strung out along the equator, mostly afloat on the tropical seas, together with floating agriculture, industries, transportation systems, and cultural facilities. And not just a few cities would be build, or a few hundred. Most likely 6,000 new cities would be build, for a million people each, together with many millions of hectares of floating farmland. The construction effort on this scale, as I said, wouldn't pose a problem either for a sane society. The technologies for it do already exist, and so do the required materials and energy resources. No miracles would be required to get this done. In a sane society, miracles are not needed, or prayed for. And the construction process itself, as an intelligent process, would require almost no effort, so that the entire construction process would be but a sideline issue, completed with a song.

With automated, high-temperature, large-scale industrial processes, almost anything can be produced, and be produced with such efficiency that the produced housing in the new cities would be provided by society to itself for free as an investment into its self-development. Once one takes the 'matter' factor out of the creative equation, what is presently unimaginable becomes routinely possible. That's the dimension of healing that can be accomplished that the "scientific statement of being" inspires that is an element of the Christ, reflected in universal love.

The above is not a dream landscape. It is not a case of looking too high, or expecting too much. It reflects merely the truly human landscape where anything less would be inhuman.

The 'matter' factor is choking us at the present. Even

money is treated as 'matter', while in real terms it is not a factor in the creative equation. Money has never built anything. Man is the builder and creator of worlds. Love for one-another, for the precious humanity that we all share, is the creative impetus with such a rich abundance that cultures flourish, beauty abounds, and scientific development steps beyond limits that are not even imagined as yet. This is the picture of the human future in a renaissance world, which is invariably a spiritual world.

When money becomes unbound from the materialistic value system, it begins to become an element in the creative spiritual process that advances the power and the quality of human living universally. Then the question will no longer be asked, "What will it cost?" The question will then become, "What else can we built and create to make our world a richer place for universal living.

On the other hand, should we fail to choose this course of deleting materialistic concepts, which is a course of defeating small-minded thinking, and would go on regarding the universe itself as made of matter, self-consuming, entropic, and dying, none of the above would happen or even be possible. In this case society would lay itself down to die, unwittingly, doing nothing to avoid the then unavoidable orgy of universal suicide by starvation that would be assured when the global agriculture begins to fail under a dimmer and colder Sun. The world is presently committed to this path that is utterly inhuman in nature, as in consequence.

When Mary Baker Eddy composed the scientific statement of being, she fully exposed what is proving to be an immensely deep-reaching issue. It is of a depth that not even the most-able of her students at the time had been able to grasp in full, if at all.

For example, it is reported in a book named Collectania (p.48), that Mary Baker Eddy once addressed her secretary Adam Dickey and touched his hand with a finger, asking, "What is this?" He is said to have replied "matter." She said to him, "it is not; it is Spirit." It is further reported that at another time she looked at him and said, "You are Spirit." He said to her, "No Mother, I am spiritual." She said to him again, in an emphatic manner, "You are Spirit." He replied that he could not see it. "She repeated, for a third time by then, "You are Spirit."

Did she overturn, with these statements, her own saying in the scientific statement of being, that man is spiritual? That's an important question, isn't it? Evidently, both statements are true, depending on where one stands. Mary Baker Eddy defined the top level of the scientific translation of mortal mind, as spiritual. At this level much of materiality has become lost. The quality of Spirit is shining through. At this point, one regards oneself as Spirit-like, or spiritual in nature. Mary Baker Eddy seems to suggest that a person or society can be healed of the last vestige of the mortal-mind trap, so that nothing remains but the translation of immortal mind to unfold as full. At this point God and man is one, just as Principle and its idea is one, and this one is Spirit.

Adam Dickey didn't see the connection. Are we prepared today to reach that far? Are we prepared to look at the Sun as Spirit, or plasma as physics have it? Are we prepared to move with it to wherever its boundless dynamics take us?

Our long history on this planet spans back more than 2 million years. As I said before, this history unfolded for more than 85% of the time in deep-cold glaciation climates. We are children of the ice ages. We came out of the last Ice Age with a sparse population of less than 10 million people worldwide, as some researchers estimate, but we came out richer. We now stand at a height of scientific and spiritual development were no one has stood before. We stand with the capability today, as the children of the divine Spirit, and face down the physical universe with a smile and say to it, where is your sting? We are greater than being merely physical, or merely spiritual. We can see the future 30 years before it happens, and can respond to it in the present. We are the idea of God, omnipotence is our second name. We have the capacity developed to create new worlds for our living, almost on demand as the dynamics of the universe require. Doesn't this qualify for the term, omnipotence? If it doesn't, what would?

The need for a fuller development of Christian Science, especially the foundation it is built on, its ACME, is greater today than it has been at any time in history while the field of Christian Science has collapsed itself by its isolation from its very center, in a form of self-denial. The collapse is not surprising. In the material ideology of the universe everything is regarded as existing in isolation, independent, disconnected, consuming itself, burning itself out. Every sun is deemed to exist in this manner, to be its own master without an active link to the whole as if it wasn't a part of the universe. This perception is typical for the manner that all material systems are deemed to operate. Ironically the Christian Science field has fallen into this materialistic trap. Every one of its major functional units is seen as an isolated creation, like one of so many books on a shelf that have no relationship to one another. Occasionally some items are quoted from them in a Bible Lesson sermon without the recognition being made what the quoted item signifies. But this isn't how the real universe functions.

In astrophysics, in the real universe, in the universe of Spirit, in the plasma universe of creative principles, nothing is isolated. Nothing exists disconnected on its own. In the real universe every sun and star, galaxy, and cluster of galaxies, is linked with streams of plasma that are a part of the universe itself. Without this universal linkage that combines the universe into one, nothing would exist. Sure, large arrays of theories are fielded that would justify the concept of a universally disconnected cosmos of independent conglomerations of matter. However, no plausible evidence exists that supports these impossible theories. All evidence points to the universe of Spirit where nothing exists in isolation. This applies to the field of Christian Science likewise, emphatically. Mary Baker Eddy created all parts of it as interlinked constructs of a single whole. When this linkage is lost, the hole is lost, and the parts are regarded as is customary in materialistic convention. In this manner the Christian Science field deprives itself and loses itself, as has indeed happened. But why should this continue? Why should the field of Christian Science stoop so low as to sing the materialistic-universe song where death is the only possible outcome according to material theory, and sing this song in all its countless forms and melodies and tones? The song has caused great damage, and now stands poised to annihilate much of humanity itself, as humanity disconnects itself from its scientific spiritual nature and denies is capacity to gleam its future by its understanding of the dynamics of the real universe, and of its creative and productive power to adjust its living accordingly, with ease.

For society to sit back in its easy chair of indifference where it denies its nature and its capacity, is a deadly convention of materialistic dreaming, whereby society becomes by default a murderer of its children. Unfortunately, this is precisely what modern society has become committed to, by its bowing to the materialistic ideology and its convention of disconnected, isolated, existing. On this platform humanity has no hope to have a future, or even a future existence.

In historic terms the breakaway from materialistic convention has never been fully achieved. When Christ and Christmas was introduced by Mary Baker Eddy, it caused such confusion that she felt prompted to withdraw it. The concept of the unity of good was still far out of sight, and sadly remains so. The ACME structure that she has built all of her achievements on, is largely deemed not to exist, and when it is recognized, is declared to be irrelevant, precisely as one would expect this according to the convention of materialistic codes.

Not only do the materialistic conventions impel society to break the unity of good, almost to the point that humanity devalues itself so deeply that its value nearly disappears in the spray, it goes further than that all the way to hailing the notion that man is a cancer on the Earth that must be cured by depopulation. Yes, these are still the banner headlines, and these banners are hailed. Even the nuclear-war posture is hailed that can achieve the depopulation in the space of a coffee break. That's the present stage of the outcome of the materialistic codes that break the unity of good and isolate its elements in a process of creeping diminishment where the most precious that humanity is, disappears in the doom of oblivion.

In the last pages of her textbook on Christian Science, Mary Baker Eddy points out that the acme of Christian Science lies in the opposite, in the principle of the unity of good. She may have created the scientific statement of being in the chapter Recapitulation of her textbook, as a warning to society not to loose sight of the principle of the unity of good.

Obviously Mary Baker Eddy wasn't aware of the solar phase shift that promises to usher in the next glaciation period in the 2050s that will render most areas outside the tropics uninhabitable, if not all of them. The technological research tools that enable the detections of the relevant evidence shift didn't exist in her time. This doesn't mean that she didn't understand the underlying principle, and wasn't aware of its extreme significance. She may have been impelled by a higher impetus than her own perception to suggest that the scientific statement of being be read at every Sunday service. This isn't a by-law requirement, but a convention she may have suggested via an article in The Christian Science Journal as a part of the Order of Services in the Mother Church, that became reprinted in the Appendix of, The Church Manual.

The statement is also critical to lay a foundation for a higher-level sense of love. While the divine Principle of "Love reflected in love" was already established in her presentation of the Lord's Prayer, the dynamics of it cannot function on the materialistic platform of universal isolation. Universal love and universal isolation are opposite concepts, obviously. In the context of "Love reflected in love" humanity is one, as God and man is one, and "Principle and its idea is one."

In the context of humanity regarding itself as a unit of one, the resulting universal love for one-another manifests the image of the tallest expression of Life, which alone the Ice Age Renaissance world, and all that goes with it, can be build on, and will be build on, and become a celebration of the precious that is the supreme manifest in humanity. On a lesser platform, likely nothing will happen; and on a materialistic platform the zero-response is virtually assured, as is presently the case. In this case 99% of humanity will vanish by its self-deprivation as it denies itself a future. Then, the remaining few that the primitive Earth will be able to support by its meager resources may form a new civilization in the next interglacial period 100,000 years later and stand at the same threshold that humanity stands at today and may not fail itself again as humanity is presently committed to fail itself by denying itself as the pinnacle of life "clothed with the Sun."

It may also be that humanity becomes inspired to put aside its materialistic shroud that isolates it into impotence and assume its role as an undivided whole, even a 'sun' in itself, and thereby live to the fullest of its divine capacity. I am convinced that this type of breakout to Truth is possible.

Time is not a factor in the process of progression, either to advance it or to hinder it. The molasses of time does not produce progression. Advanced ideas do this. Nor is time an inertia that can hinder the dynamic expression of Spirit in man. Only ignorance and self-denial have this effect.

In the spiritual universe of divine Science, it is possible to lay aside ages of ignorance as if they never existed and assume positions in Truth that are rightfully ours. That's the divine way. It is our way. It is the way in the ACME of Christian Science. I am confident that we will take up the chance that is divinely ours and make it so. Then watch out! Then we will experience what has not been experienced before. We will experience God reflected in omnipotence. Thus I am convinced that we will built the needed 6,000 new cities on this platform, as a starting 'gesture,' and that we will provide them to one another for free as a natural expression of our humanity. Wars will fall by the wayside and nuclear war will never happen. And of course, the depopulation of the Earth will not be carried out then, but be forgotten. We will see ourselves as the bearer of an infinite future, and we will "make it so." We have the tools for this future already on hand.

The 3-part Apocalypse

Mary Baker Eddy's textbook chapter, The Apocalypse, consists of 3 parts. Each part represents a distinct aspect of the whole. The first is centered on the tenth chapter of Revelation that speaks of a mighty angel clothed with a cloud. Mary Baker Eddy tells us that the angel represents divine science that is at first obscure and abstract until it becomes fairly understood. The angel bears an open book for all to read and understand. But who makes the effort, or even dares? Mary Baker Eddy's contribution to the ACME structure remains largely unknown to the present day.

The second part of the chapter The Apocalypse is focused on Revelation 12. It begins with the vision of the angel clothed with the Sun, with the moon under her feet and on her head a crown of twelve stars. She was with child, but there was war in heaven. A great red dragon stood before the woman to devour her child as soon as it was born. But the child was caught up unto God and the woman fled into the wilderness where she had a place prepared for her by God. Still the dragon did not relent. There was war in heaven. Angels fought against the dragon, who prevailed not, but was cast out, and his dark angels were cast out with him. When the dragon saw that he was cast out, he prosecuted the woman, but the woman was given two wings, with which she escaped once more into the wilderness. Here a serpent appeared, a different type of dragon that issued a flood streaming from its mouth that the woman might be carried away in the flood. But the earth opened up and swallowed the flood.

Mary Baker Eddy tells us that the woman represents generic man, the spiritual idea of God, clothed with the radiance of spiritual Truth, who has put all matter under her feet, and her child is the Christ.

The type of contest presented here is interesting. While there is never war in heaven, in the realm of immutable harmony, the effect that Truth has on a small-minded humanity, on the scene of mortal mind's conventions in materiality, is that of a lion roaring that wakes somebody up.

The process is illustrated in the ACME structure. There the 3-part Apocalypse represents the concepts of Soul Truth and Love that gained special meaning by them being not defined in the Glossary, but are defined in the larger context. In this context the war scene is centered on Truth, the first row, the SCIENTIFIC TRANSLATION OF IMMORTAL MIND. The light of Truth is not isolated there, but reaches to the deepest levels of error and related materialistic conventions that errors invite. In this war (on error) the 8 opposite dual definitions come into play that give their Glossary term both a materialistic meaning and a divine significance, such as in the case of the term, Zion.

Spiritual foundation and superstructure; inspiration; spiritual strength.

Emptiness; unfaithfulness; desolation.

Yes, there is 'war' in heaven, like the movement of a perpetual fresh wind that clears the smog away. In this context the Earth - the hard evidence of the universe of Spirit reflected in the physical plasma universe – will help humanity. Such hard evidence, which exists plentiful now, will swallow up the floods of opposing materialistic, small-minded notions.

The third part of the three-part chapter, The Apocalypse, is focused on Revelation 21 that presents the city four square. It deals with the Christ, the certainty of healing, a new heaven and a new earth, and the former having past away. Love reflected in love is the Christ-process that heals all wounds. Love is the third term that is not defined in the Glossary, but is defined by the larger context. The Christ wages its own war on the lower platform, as in heaven. That's where the 8 soft vertical dual definitions in the Glossary structures come into play, reaching down to the third level from the second, uplifting Christianity spiritually, from where it can raise itself with universal love to the first level.

While the book of Revelation in the Bible presents a sequence of metaphoric events, in real terms all aspects are integrated in the 4-square ACME structure as timeless realities, all happening simultaneously. In this simultaneity of eternity, lies the certainty of healing, which simply means, a comprehensive end of all evil, where evil vanishes as if it never existed, which indeed, is the case.

With this in mind I would say that the next Ice Age will happen, as ice ages always have according to the dynamics of the universe, and that our response to it this time, will be on a higher level that has never been achieved before. And so I would like to celebrate the certainty that we will build 6,000 new cities for starters, in the next 30 years, and create a new world and go on from there. Yes, there will be war in heaven along the path that impels us to stir our stumps, to get on with it. This means that we will come out richer by this experience than in our tallest dreams. John saw the principles involved, in operation. Mary Baker Eddy saw that John was right, and that what he saw is the "acme" of Christian Science. Thus, she built on it.

For supporting publications and video presentations related to plasma physics and the coming Ice Age, please see my extensive illustrated science project, **"Cool Science for Kids to have a Future in the Near Ice Age World."** at: http://www.ice-age-ahead-iaa.ca/

The Lodging for the Rose

12 novels and the poem

Harvest is Seedtime

By Rolf A. F. Witzsche

Development-type structures

3	6	9	12
2	5	8	11
1	4	7	10

Seeds are we, wind-blown

Carriers of a secret still unknown

Poems in the words of nature

Sentinels of an Intelligence yet unseen

Prophets of the enduring

Apostles in an endless landscape

4000

Discovering Love

Seeds, wind-blown
Poems in the words of nature
Sentinels of an Intelligence yet unseen

the novel Discovering Love

Apostles in an endless landscape

Harvest is seedtime, thoughts ripening

Carried as by a great wind

Carriers of secrets to unfold

Thoughts winged with Purpose

A force waiting, silent

Patiently waiting for the moment

3000

Harvest is seedtime, thoughts ripening
Carried as by a great wind
Carriers of secrets to unfold

The Ice Age
Challenge

the novel The Ice Age Challenge

Patiently waiting for the moment

Thoughts do awaken
Roused by the moist warmth in spring
Cascades of colors, colors of life
Bright yellows, bursts of silver-white
Thoughts becoming creations
Monuments of genius, builders of worlds

2000

**Roses at Dawn
in Ice Age World**

Cascades of colors,
colors of life
Bright yellows,
bursts of silver-white
Thoughts becoming creations

the novel Roses as Dawn in an Ice Age World

Monuments of genius, builders of worlds

Who owns the seeds? Do we?

Who can fathom their wonder?

Life flows from then in great rivers

Rivers trailing into oceans

In them we are alone, each one is alone

Each thought is sovereign, beauty is its song

o4oo

Winning without Victory

Who owns the seeds? Do we?
Life flows from them in great rivers
Rivers trailing into oceans

the novel Winning without Victory

Each thought is sovereign, beauty is its song

Thoughts are seeds, becoming ideas

Alive in discovering

Alive in listening

Alive in being touched by love

Alive in loving

Alive...

o3oo

the novel Seascapes and Sand

Alive...

Like seeds, thoughts fall to the ground

Potentials are lost

Hard grounds kill the precious

But we are Man

Hard ground becomes tilled, watered

The precious is nurtured in loving

o2oo

Hard grounds kill the precious
But we are Man
Hard ground becomes tilled, watered

The Flat Earth Society

PORSCHE

the novel The Flat Earth Society

The precious is nurtured in loving

Love for one-another, the human spring
Mankind is afloat in a sea that is Love
Seeds germinate, become plants
Roots break the ground
Love lifts the barriers, patiently
Silently waiting, reaching for the sky

oo4o

Roots break the ground
Love lifts the barriers, patiently
Silently waiting, reaching for the sky

Glass Barriers

the novel Glass Barriers

Silently waiting, reaching for the sky

Thoughts are the Universe unfolding
Landscapes of brilliance, ideas of power
Substance for enriching one-another
Substance of the forever maturing
Thoughts bearing new seeds within
Seeds for splendors beyond dreams

oo3o

Landscapes of brilliance, ideas of power
Substance for enriching one-another
Substance of the forever maturing

Coffee, Sex,
and Biscuits

the novel Coffee, Sex, and Biscuits

Seeds for splendors beyond dreams

Each harvest is seedtime

A seed becomes a plant bearing new seeds

A thought unfolding, bears up civilization

A spark in the heart, bears the 'fire' of life

New worlds are created in the 'fire' of passion

We are the bearers of a 'fire' that is light

oo2o

Endless Horizons

A thought unfolding,
  bears up civilization
A spark in the heart,
  bears the 'fire' of life
New worlds are created
  in the 'fire' of passion

the novel Endless Horizons

We are the bearers of a 'fire' that is light

Builders of worlds are we
New Worlds, which have never been
Precious with riches grander than our own
Nature is Love reflected in loving
Love paints with the colors of its endless spring
Love paints us all - but who owns the seed?

ooo4

Precious with riches
grander than our own
Nature is Love
reflected in loving
Love paints with the colors
of its endless spring

Angels
of Sex
in Queensland

the novel Angels of Sex in Queensland

Love paints us all - but who owns the seed?

Who owns the cradle for the seed?
Name it Intelligence, name it the Universe
Thoughts are seeds from an infinite fountain
Monuments of grandeur of good
Fields of flowers dancing in the sunshine
All nature whispers this to us

ooo3

Thoughts are seeds from an infinite fountain
Monuments of grandeur of good
Fields of flowers dancing in the sunshine

Sword of Aquarius

the novel Sword of Aquarius

All nature whispers this to us

The melody of nature - what a song!

Whispers of a splendor grander than the heavens

Like seeds are we - we whisper too

Seeds bearing gifts for the world

Gifts wrapped up in sunshine

Gems are we - unfolding a majestic song!

ooo2

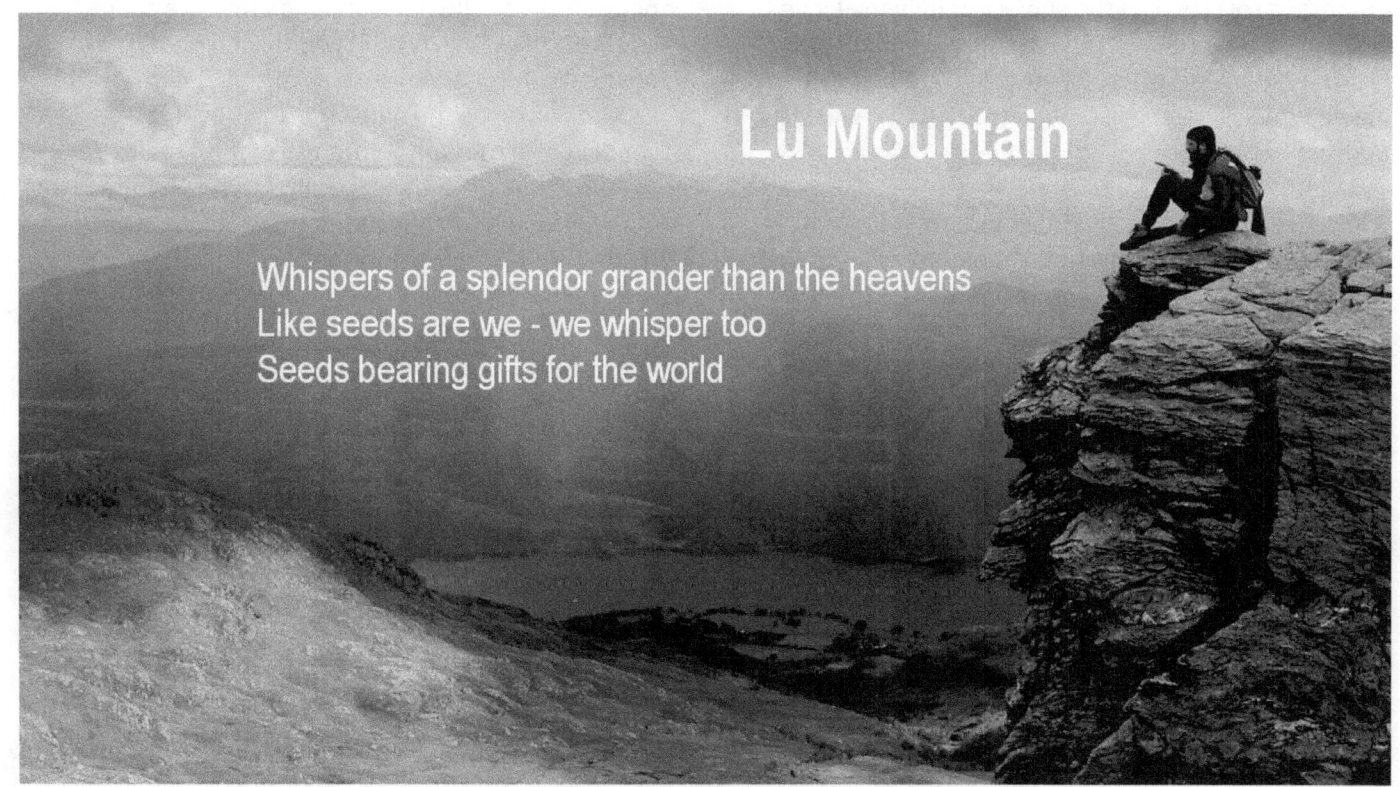

Lu Mountain

Whispers of a splendor grander than the heavens
Like seeds are we - we whisper too
Seeds bearing gifts for the world

the novel Lu Mountain

Gems are we - unfolding a majestic song!

The above 12 novels, of the series, *"The Lodging for the Rose,"* present together an epic tale that is centered on the monumental question, "What is the Principle of universal love?" The question is explored in the context of the social domain, at the grass-roots level where "the rubber meets the road," so to speak. The exploration is challenging, of course, in a small-minded world where society lives almost universally divided, isolated, and where everything is privatized, diminished, and to a large degree destroyed, or is prepared to be destroyed.

The series is large, as it needs to be as a means for exploring such a large subject as universal love, or **"Love reflected in love,"** which is evidently not a trivial issue, because divine Love is the universal impetus that humanity has forever in its heart to respond to, rather than being a "catch me if you can" private emotional issue, surrounded by fences of numerous sorts. Nor is the subject of universal love small in its effect. It will ultimately determine the future of mankind when the next Ice Age begins that may be only 30 years distant, according to a wide body of evidence.

If love remains small, privatized and isolated, as it typically is at the present time, humanity will not be able to build 6,000 new cities in less than 30 years that it requires in order to relocate itself into the tropics where its living would be able to continue and flourish in an Ice Age world, while otherwise 99% of the people of the world would perish for the lack of food when the regions outside the tropics become uninhabitable.

However, with divine Love being reflected in universal love in every respect, the relocating of 6 billion people, together with their agriculture and industries, wouldn't pose a problem. The physical resources for this to happen do all exist. Indeed, all of this may yet happen, as unlikely as it may seem in the current world. With the entire series, The Lodging for the Rose, unfolding in the lower three

rows of the ACME structure, the operating impetus is the Christ, which Mary Baker Eddy has associated with the synonym for God, **divine Love.** This too, is a part of the ACME of Christian Science. The healing process has already been defined by Mary Baker Eddy in the TRANSLATION OF MORTAL MIND out of its petty smallness, with the Christ, towards society's participation in the TRANSLATION OF IMMORTAL MIND.

While this is big, my poem doesn't end here. One verse remains. It has two novels attached to it that are not a part on the 12-volume series.

The first of these, **the novel, *Flight Without Limits*,** is focused on the quality of Truth. The novel is focused on an ancient legend rooted in North American native West Coast culture. The legend speaks of a magic canoe that can take a traveler instantly to any place desired. The novel extends this to a space voyage in various ways. While this technology is physically impossible, such as to traverse the vast extends of space in an instant, the novel takes the concept as a metaphor into the mental realm where we should all have the capability to reach for aspects of divine Truth in an instant and experience their reality. Time and distance are evidently not factors in the realm of God.

The second novel that is not a part of the *Lodging for the Rose*, **the novel, *Brighter than the Sun*,** is focused on Love with a wide horizon in the context of nuclear war. Numerous factors cause an accident to happen in the novel, which causes immense damage. The novel explores in a 6-stage story the dimensions in loving that could have prevented the accident in the first place, which a fuller love might have inspired if the 'technology' for it had existed in the beginning. In the ACME of Christian Science, Love being reflected in love, truly has the capacity to be **brighter than the Sun. All healing in the world depends on love, actively reflecting divine Love.** More than that we cannot experience.

Listen to the song

Listen to the heart

Listen to the silence where strands of love unfold

the novel Flight Without Limits

...where strands of love unfold

Listen to the symphony of our humanity
In this symphony we are One
One with the Universe itself.

the novel Brighter than the Sun

One with one-another...

The novels presented here are available in print and in "kindle" tablet form, see the list below. The novels are further available for free in PDF form, in audio form, and in online form, on the website:
http://www.ice-age-ahead-iaa.ca/

Discovering Love https://www.createspace.com/6020651 http://www.amazon.ca/dp/B01AX32O3E	**Coffee Sex and Biscuits** https://www.createspace.com/6024702 http://www.amazon.ca/dp/B01AYSPMPA
The Ice Age Challenge https://www.createspace.com/6022783 http://www.amazon.ca/dp/B01AYOPD7G	**Endless Horizons** https://www.createspace.com/6024784 http://www.amazon.ca/dp/B01AYSPMPK
Roses at Dawn in an Ice Age World https://www.createspace.com/6022816 http://www.amazon.ca/dp/B01AYOPD8U	**Angels of Sex in Queensland** https://www.createspace.com/6024827 http://www.amazon.ca/dp/B01AYSPMXC
Winning Without Victory https://www.createspace.com/6024438 http://www.amazon.ca/dp/B01AYSPMSM	**Sword of Aquarius** https://www.createspace.com/6024872 http://www.amazon.ca/dp/B01AYNKLQU
Seascapes and Sand https://www.createspace.com/6024503 http://www.amazon.ca/dp/B01AYSPMR8	**Lu Mountain** https://www.createspace.com/6024903 http://www.amazon.ca/dp/B01AYNXQZI
The Flat Earth Society https://www.createspace.com/6024557 http://www.amazon.ca/dp/B01AYSPN04	**Flight Without Limits** https://www.createspace.com/6019521 http://www.amazon.ca/dp/B01AR556UG
Glass Barriers https://www.createspace.com/6024647 http://www.amazon.ca/dp/B01AYSPMY6	**Brighter than the Sun** https://www.createspace.com/6018971 http://www.amazon.ca/dp/B01B3E0Y0C

The Appendix

Major items from the Christian Science textbook,
Science and Health with Key to the Scriptures by Mary Baker Eddy
which are referenced in this book, item by item

1. The Platform of Christian Science
2. The Chapter Recapitulation
3. The Glossary
4. The Manual of the Mother Church

The Platform of Christian Science

Written by Mary Baker Eddy

That which was from the beginning, which we have heard, which we have seen with our eyes, which we have looked upon, and our hands have handled, of the Word of life, . . . That which we have seen and heard declare we unto you, that ye also may have fellowship with us: and truly our fellowship is with the Father, and with His Son Jesus Christ.
- John, First Epistle.

Here I stand. I can do no otherwise; so help me God! Amen!
- Martin Luther.

When the following platform is understood and the letter and the spirit bear witness, the infallibility of divine metaphysics will be demonstrated.

ooo1

-1- The deific supremacy 330

I. God is infinite, the only Life, substance, Spirit, or Soul, the only intelligence of the universe, including man. Eye hath neither seen God nor His image and likeness. Neither God nor the perfect man can be discerned by the material senses. The individuality of Spirit, or the infinite, is unknown, and thus a knowledge of it is left either to human conjecture or to the revelation of divine Science.

-2- The deific definitions 330

II. God is what the Scriptures declare Him to be, - Life, Truth, Love. Spirit is divine Principle, and divine Principle is Love, and Love is Mind, and Mind is not both good and bad, for God is Mind; therefore there is in reality one Mind only, because there is one God.

oo1o

-3- Evil obsolete 330

III. The notion that both evil and good are real is a delusion of material sense, which Science annihilates. Evil is nothing, no thing, mind, nor power. As manifested by mankind it stands for a lie, nothing claiming to be something, - for lust, dishonesty, selfishness, envy, hypocrisy, slander, hate, theft, adultery, murder, dementia, insanity, inanity, devil, hell, with all the etceteras that word includes.

-4- Life the creator 331

IV. God is divine Life, and Life is no more confined to the forms which reflect it than substance is in its shadow. If life were in mortal man or material things, it would be subject to their limitations and would end in death. Life is Mind, the creator reflected in His creations. If He dwelt within what He creates, God would not be reflected but absorbed, and the Science of being would be forever lost through a mortal sense, which falsely testifies to a beginning and an end.

o1oo

-5- Allness of Spirit 331

V. The Scriptures imply that God is All-in-all. From this it follows that nothing possesses reality nor existence except the divine Mind and His ideas. The Scriptures also declare that God is Spirit. Therefore in Spirit all is harmony, and there can be no discord; all is Life, and there is no death. Everything in God's universe expresses Him.

-6- The universal cause 331

VI. God is individual, incorporeal. He is divine Principle, Love, the universal cause, the only creator, and there is no other self-existence. He is all-inclusive, and is reflected by all that is real and eternal and by nothing else. He fills all space, and it is impossible to conceive of such omnipresence and individuality except as infinite Spirit or Mind. Hence all is Spirit and spiritual.

1ooo

-7- Divine trinity 331

VII. Life, Truth, and Love constitute the triune Person called God, - that is, the triply divine Principle, Love. They represent a trinity in unity, three in one, - the same in essence, though multiform in office: God the Father-Mother; Christ the spiritual idea of sonship; divine Science or the Holy Comforter. These three express in divine Science the threefold, essential nature of the infinite. They also indicate the divine Principle of scientific being, the intelligent relation of God to man and the universe.

-8- Father-Mother 332

VIII. Father-Mother is the name for Deity, which indicates His tender relationship to His spiritual creation. As the apostle expressed it in words which he quoted with approbation from a classic poet: "For we are also His offspring."

ooo2

-9- The Son of God 332

IX. Jesus was born of Mary. Christ is the true idea voicing good, the divine message from God to men speaking to the human consciousness. The Christ is incorporeal, spiritual, - yea, the divine image and likeness, dispelling the illusions of the senses; the Way, the Truth, and the Life, healing the sick and casting out evils, destroying sin, disease, and death. As Paul says: "There is one God, and one mediator between God and men, the man Christ Jesus." The corporeal man Jesus was human.

-10- Holy Ghost or Comforter 332

X. Jesus demonstrated Christ; he proved that Christ is the divine idea of God - the Holy Ghost, or Comforter, revealing the divine Principle, Love, and leading into all truth.

oo2o

-11- Christ Jesus 332

XI. Jesus was the son of a virgin. He was appointed to speak God's word and to appear to mortals in such a form of humanity as they could understand as well as perceive. Mary's conception of him was spiritual, for only purity could reflect Truth and Love, which were plainly incarnate in the good and pure Christ Jesus. He expressed the highest type of divinity, which a fleshly form could express in that age. Into the real and ideal man the fleshly element cannot enter. Thus it is that Christ illustrates the coincidence, or spiritual agreement, between God and man in His image.

-12- Messiah or Christ 333

XII. The word Christ is not properly a synonym for Jesus, though it is commonly so used. Jesus was a human name, which belonged to him in common with other Hebrew boys and men, for it is identical with the name Joshua, the renowned Hebrew leader. On the other hand, Christ is not a name so much as the divine title of Jesus. Christ expresses God's spiritual, eternal nature. The name is synonymous with Messiah, and alludes to the spirituality which is taught, illustrated, and demonstrated in the life of which Christ Jesus was the embodiment. The proper name of our Master in the Greek was Jesus the Christ; but Christ Jesus better signifies the Godlike.

-13- The divine Principle and idea 333

XIII. The advent of Jesus of Nazareth marked the first century of the Christian era, but the Christ is without beginning of years or end of days. Throughout all generations both before and after the Christian era, the Christ, as the spiritual idea, - the reflection of God, - has come with some measure of power and grace to all prepared to receive Christ, Truth. Abraham, Jacob, Moses, and the prophets caught glorious glimpses of the Messiah, or Christ, which baptized these seers in the divine nature, the essence of Love. The divine image, idea, or Christ was, is, and ever will be inseparable from the divine Principle, God. Jesus referred to this unity of his spiritual identity thus: "Before Abraham was, I am;" "I and my Father are one;" "My Father is greater than I." The one Spirit includes all identities.

-14- Spiritual oneness 333

XIV. By these sayings Jesus meant, not that the human Jesus was or is eternal, but that the divine idea or Christ was and is so and therefore antedated Abraham; not that the corporeal Jesus was one with the Father, but that the spiritual idea, Christ, dwells forever in the bosom of the Father, God, from which it illumines heaven and earth; not that the Father is greater than Spirit, which is God, but greater, infinitely greater, than the fleshly Jesus, whose earthly career was brief.

-15- The Son's duality 334

XV. The invisible Christ was imperceptible to the so-called personal senses, whereas Jesus appeared as a bodily existence. This dual personality of the unseen and the seen, the spiritual and material, the eternal Christ and the corporeal Jesus manifest in flesh, continued until the Master's ascension, when the human, material concept, or Jesus, disappeared, while the spiritual self, or Christ, continues to exist in the eternal order of divine Science, taking away the sins of the world, as the Christ has always done, even before the human Jesus was incarnate to mortal eyes.

-16- Eternity of the Christ 334

XVI. This was "the Lamb slain from the foundation of the world," - slain, that is, according to the testimony of the corporeal senses, but undying in the deific Mind. The Revelator represents the Son of man as saying (Revelation i. 17, 18): "I am the first and the last: I am he that liveth, and was dead not understood]; and, behold, I am alive for evermore, Science has explained me]." This is a mystical statement of the eternity of the Christ, and is also a reference to the human sense of Jesus crucified.

ooo3

-17- Infinite Spirit 334

XVII. Spirit being God, there is but one Spirit, for there can be but one infinite and therefore one God. There are neither spirits many nor gods many. There is no evil in Spirit, because God is Spirit. The theory, that Spirit is distinct from matter but must pass through it, or into it, to be individualized, would reduce God to dependency on matter, and establish a basis for pantheism.

-18- The only substance 335

XVIII. Spirit, God, has created all in and of Himself. Spirit never created matter. There is nothing in Spirit out of which matter could be made, for, as the Bible declares, without the Logos, the Aeon or Word of God, "was not anything made that was made." Spirit is the only substance, the invisible and indivisible infinite God. Things spiritual and eternal are substantial. Things material and temporal are insubstantial.

oo3o

-19- Soul and Spirit one 335

XIX. Soul and Spirit being one, God and Soul are one, and this one never included in a limited mind or a limited body. Spirit is eternal, divine. Nothing but Spirit, Soul, can evolve Life, for Spirit is more than all else. Because Soul is immortal, it does not exist in mortality. Soul must be incorporeal to be Spirit, for Spirit is not finite. Only by losing the false sense of Soul can we gain the eternal unfolding of Life as immortality brought to light.

-20- The one divine Mind 335

XX. Mind is the divine Principle, Love, and can produce nothing unlike the eternal Father-Mother, God. Reality is spiritual, harmonious, immutable, immortal, divine, eternal. Nothing unspiritual can be real, harmonious, or eternal. Sin, sickness, and mortality are the suppositional antipodes of Spirit, and must be contradictions of reality.

o3oo

-21- The divine Ego 335

XXI. The Ego is deathless and limitless, for limits would imply and impose ignorance. Mind is the I AM, or infinity. Mind never enters the finite. Intelligence never passes into non-intelligence, or matter. Good never enters into evil, the unlimited into the limited, the eternal into the temporal, nor the immortal into mortality. The divine Ego, or individuality, is reflected in all spiritual individuality from the infinitesimal to the infinite.

-22- The real manhood 336

XXII. Immortal man was and is God's image or idea, even the infinite expression of infinite Mind, and immortal man is coexistent and coeternal with that Mind. He has been forever in the eternal Mind, God; but infinite Mind can never be in man, but is reflected by man. The spiritual man's consciousness and individuality are reflections of God. They are the emanations of Him who is Life, Truth, and Love. Immortal man is not and never was material, but always spiritual and eternal.

3ooo

-23- Indivisibility of the infinite 336

XXIII. God is indivisible. A portion of God could not enter man; neither could God's fulness be reflected by a single man, else God would be manifestly finite, lose the deific character, and become less than God. Allness is the measure of the infinite, and nothing less can express God.

-24- God the parent Mind 336

XXIV. God, the divine Principle of man, and man in God's likeness are inseparable, harmonious, and eternal. The Science of being furnishes the rule of perfection, and brings immortality to light. God and man are not the same, but in the order of divine Science, God and man coexist and are eternal. God is the parent Mind, and man is God's spiritual offspring.

0004

-25- Man reflects the perfect God 336

XXV. God is individual and personal in a scientific sense, but not in any anthropomorphic sense. Therefore man, reflecting God, cannot lose his individuality; but as material sensation, or a soul in the body, blind mortals do lose sight of spiritual individuality. Material personality is not realism; it is not the reflection or likeness of Spirit, the perfect God. Sensualism is not bliss, but bondage. For true happiness, man must harmonize with his Principle, divine Love; the Son must be in accord with the Father, in conformity with Christ. According to divine Science, man is in a degree as perfect as the Mind that forms him. The truth of being makes man harmonious and immortal, while error is mortal and discordant.

-26- Purity the path to perfection 337

XXVI. Christian Science demonstrates that none but the pure in heart can see God, as the gospel teaches. In proportion to his purity is man perfect; and perfection is the order of celestial being which demonstrates Life in Christ, Life's spiritual ideal.

---
oo4o

-27- True idea of man 337

XXVII. The true idea of man, as the reflection of the invisible God, is as incomprehensible to the limited senses as is man's infinite Principle. The visible universe and material man are the poor counterfeits of the invisible universe and spiritual man. Eternal things (verities) are God's thoughts as they exist in the spiritual realm of the real. Temporal things are the thoughts of mortals and are the unreal, being the opposite of the real or the spiritual and eternal.

-28- Truth demonstrated 337

XXVIII. Subject sickness, sin, and death to the rule of health and holiness in Christian Science, and you ascertain that this Science is demonstrably true, for it heals the sick and sinning as no other system can. Christian Science, rightly understood, leads to eternal harmony. It brings to light the only living and true God and man as made in His likeness; whereas the opposite belief - that man originates in matter and has beginning and end, that he is both soul and body, both good and evil, both spiritual and material - terminates in discord and mortality, in the error which must be destroyed by Truth. The mortality of material man proves that error has been ingrafted into the premises and conclusions of material and mortal humanity.

---
o4oo

-29- Adam not ideal man 338

XXIX. The word Adam is from the Hebrew adamah, signifying the red color of the ground, dust, nothingness. Divide the name Adam into two syllables, and it reads, a dam, or obstruction. This suggests the thought of something fluid, of mortal mind in solution. It further suggests the thought of that "darkness . . . upon the face of the deep," when matter or dust was deemed the agent of Deity in creating man, - when matter, as that which is accursed, stood opposed to Spirit. Here a dam is not a mere play upon words; it stands for obstruction, error, even the supposed separation of man from God, and the obstacle which the serpent, sin, would impose between man and his creator. The dissection and definition of words, aside from their metaphysical derivation, is not scientific. Jehovah declared the ground was accursed; and from this ground, or matter, sprang Adam, notwithstanding God had blessed the earth "for man's sake." From this it follows that Adam was not the ideal man for whom the earth was blessed. The ideal man was revealed in due time, and was known as Christ Jesus.

-30- Divine pardon 339

XXX. The destruction of sin is the divine method of pardon. Divine Life destroys death, Truth destroys error, and Love destroys hate. Being destroyed, sin needs no other form of forgiveness. Does not God's pardon, destroying any one sin, prophesy and involve the final destruction of all sin?

4000

-31- Evil not produced by God 339

XXXI. Since God is All, there is no room for His unlikeness. God, Spirit, alone created all, and called it good. Therefore evil, being contrary to good, is unreal, and cannot be the product of God. A sinner can receive no encouragement from the fact that Science demonstrates the unreality of evil, for the sinner would make a reality of sin, - would make that real which is unreal, and thus heap up "wrath against the day of wrath." He is joining in a conspiracy against himself, - against his own awakening to the awful unreality by which he has been deceived. Only those, who repent of sin and forsake the unreal, can fully understand the unreality of evil.

-32- Basis of health and immortality 339

XXXII. As the mythology of pagan Rome has yielded to a more spiritual idea of Deity, so will our material theories yield to spiritual ideas, until the finite gives place to the infinite, sickness to health, sin to holiness, and God's kingdom comes "in earth, as it is in heaven." The basis of all health, sinlessness, and immortality is the great fact that God is the only Mind; and this Mind must be not merely believed, but it must be understood. To get rid of sin through Science, is to divest sin of any supposed mind or reality, and never to admit that sin can have intelligence or power, pain or pleasure. You conquer error by denying its verity. Our various theories will never lose their imaginary power for good or evil, until we lose our faith in them and make life its own proof of harmony and God.

-Platform summation- 340

This text in the book of Ecclesiastes conveys the Christian Science thought, especially when the word **duty**, which is not in the original, is omitted: "Let us hear the conclusion of the whole matter: Fear God, and keep His commandments: for this is the whole duty of man." In other words: Let us hear the conclusion of the whole matter: love God and keep His commandments: for this is the whole of man in His image and likeness. Divine Love is infinite. Therefore all that really exists is in and of God, and manifests His love.

"Thou shalt have no other gods before me." (Exodus xx. 3.) The First Commandment is my favorite text. It demonstrates Christian Science. It inculcates the tri-unity of God, Spirit, Mind; it signifies that man shall have no other spirit or mind but God, eternal good, and that all men shall have one Mind. The divine Principle of the First Commandment bases the Science of being, by which man demonstrates health, holiness, and life eternal. One infinite God, good, unifies men and nations; constitutes the brotherhood of man; ends wars; fulfils the Scripture, "Love thy neighbor as thyself;" annihilates pagan and Christian idolatry, - whatever is wrong in social, civil, criminal, political, and religious codes; equalizes the sexes; annuls the curse on man, and leaves nothing that can sin, suffer, be punished or destroyed.

The Chapter Recapitulation

Editor's note. For reference purposes, the Recapitulation questions have been sequentially numbered, and the sequentially applicable elements from the definition for the term Adam have been added, together with the sequentially applicable Bible Lesson topic. The original page numbers have applied behind the subheadings.

Written by Mary Baker Eddy

For precept must be upon precept, precept upon precept; line upon line, line upon line; here a little, and there a little. - Isaiah.

This chapter (the chapter Recapitulation) is from the first edition of the author's class-book, copyrighted in 1870. After much labor and increased spiritual understanding, she revised that treatise for this volume in 1875. Absolute Christian Science pervades its statements, to elucidate scientific metaphysics.

QUESTIONS AND ANSWERS

ooo2

~ Question 1. - **What is God?**~
~Adam: Error~
~Lesson topic: GOD~

Answer. - God is incorporeal, divine, supreme, infinite Mind, Spirit, Soul, Principle, Life, Truth, Love.

~ Question 2. - **Are these terms synonymous?**~
~Adam: A falsity ~
~Lesson topic: SACRAMENT~

Answer. - They are. They refer to one absolute God. They are also intended to express the nature, essence, and wholeness of Deity. The attributes of God are justice, mercy, wisdom, goodness, and so on.

oo2o

~ Question 3. - Is there more than one God or Principle?~

~Adam: The belief in "original sin," sickness, and death.~
~Lesson topic: LIFE~

Answer. - There is not. Principle and its idea is one, and this one is God, omnipotent, omniscient, and omnipresent Being, and His reflection is man and the universe. Omni is adopted from the Latin adjective signifying all. Hence God combines all-power or potency, all-science or true knowledge, all-presence. The varied manifestations of Christian Science indicate Mind, never matter, and have one Principle.

~ Question 4. - What are spirits and souls?~

~Adam: Evil~
~Lesson topic: TRUTH~

Answer. -

(Real versus unreal)
To human belief, they are personalities constituted of mind and matter, life and death, truth and error, good and evil; but these contrasting pairs of terms represent contraries, as Christian Science reveals, which neither dwell together nor assimilate. Truth is immortal; error is mortal. Truth is limitless; error is limited. Truth is intelligent; error is non-intelligent. Moreover, Truth is real, and error is unreal. This last statement contains the point you will most reluctantly admit, although first and last it is the most important to understand.

Mankind redeemed 466

The term souls or spirits is as improper as the term gods. Soul or Spirit signifies Deity and nothing else. There is no finite soul nor spirit. Soul or Spirit means only one Mind, and cannot be rendered in the plural. Heathen mythology and Jewish theology have perpetuated the fallacy that intelligence, soul, and life can be in matter; and idolatry and ritualism are the outcome of all man-made beliefs. The Science of Christianity comes with fan in hand to separate the chaff from the wheat. Science will declare God aright, and Christianity will demonstrate this declaration and its divine Principle, making mankind better physically, morally, and spiritually.

o2oo

~ Question 5. - **What are the demands of the Science of Soul?**
~Adam: The opposite of good,--of God and His creation.~
~Lesson topic: LOVE~

Answer. -
(Two chief commands)

The first demand of this Science is, "Thou shalt have no other gods before me." This me is Spirit. Therefore the command means this: Thou shalt have no intelligence, no life, no substance, no truth, no love, but that which is spiritual. The second is like unto it, "Thou shalt love thy neighbor as thyself." It should be thoroughly understood that all men have one Mind, one God and Father, one Life, Truth, and Love. Mankind will become perfect in proportion as this fact becomes apparent, war will cease and the true brotherhood of man will be established. Having no other gods, turning to no other but the one perfect Mind to guide him, man is the likeness of God, pure and eternal, having that Mind which was also in Christ.

Soul not confined in body 467

Science reveals Spirit, Soul, as not in the body, and God as not in man but as reflected by man. The greater cannot be in the lesser. The belief that the greater can be in the lesser is an error that works ill. This is a leading point in the Science of Soul, that Principle is not in its idea. Spirit, Soul, is not confined in man, and is never in matter. We reason imperfectly from effect to cause, when we conclude that matter is the effect of Spirit; but a priori reasoning shows material existence to be enigmatical. Spirit gives the true mental idea. We cannot interpret Spirit, Mind, through matter. Matter neither sees, hears, nor feels.

Reasoning from cause to effect in the Science of Mind, we begin with Mind, which must be understood through the idea which expresses it and cannot be learned from its opposite, matter. Thus we arrive at Truth, or intelligence, which evolves its own unerring idea and never can be coordinate with human illusions. If Soul sinned, it would be mortal, for sin is mortality's self, because it kills itself. If Truth is immortal, error must be mortal, because error is unlike Truth. Because Soul is immortal, Soul cannot sin, for sin is not the eternal verity of being.

~ Question 6. - **What is the scientific statement of being?**~
~Adam: A curse~
~Lesson topic: SPIRIT~

Answer. - There is no life, truth, intelligence, nor substance in matter. All is infinite Mind and its infinite manifestation, for God is All-in-all. Spirit is immortal Truth; matter is mortal error. Spirit is the real and eternal; matter is the unreal and temporal. Spirit is God, and man is His image and likeness. Therefore man is not material; he is spiritual.

2ooo

~ Question 7. - **What is substance?**~

~Adam: A belief in intelligent matter, finiteness, and mortality.~
~Lesson topic: SOUL~

Answer. -

(Spiritual synonyms)

Substance is that which is eternal and incapable of discord and decay. Truth, Life, and Love are substance, as the Scriptures use this word in Hebrews: "The substance of things hoped for, the evidence of things not seen." Spirit, the synonym of Mind, Soul, or God, is the only real substance. The spiritual universe, including individual man, is a compound idea, reflecting the divine substance of Spirit.

~ Question 8. - **What is Life?**~

~Adam: "Dust to dust"~
~Lesson topic: MIND~

Answer. -

(Eternity of Life)

Life is divine Principle, Mind, Soul, Spirit. Life is without beginning and without end. Eternity, not time, expresses the thought of Life, and time is no part of eternity. One ceases in proportion as the other is recognized. Time is finite; eternity is forever infinite. Life is neither in nor of matter. What is termed matter is unknown to Spirit, which includes in itself all substance and is Life eternal. Matter is a human concept. Life is divine Mind. Life is not limited. Death and finiteness are unknown to Life. If Life ever had a beginning, it would also have an ending.

0oo3

~ Question 9. - **What is intelligence?**~

~Adam: Red sandstone.~
~Lesson topic: CHRIST JESUS~

Answer. - Intelligence is omniscience, omnipresence, and omnipotence. It is the primal and

eternal quality of infinite Mind, of the triune Principle, - Life, Truth, and Love, - named God.

~ Question 10. - What is Mind?~
~Adam: Nothingness.~
~Lesson topic: MAN~

Answer. -
(True sense of infinitude)

Mind is God. The exterminator of error is the great truth that God, good, is the only Mind, and that the supposititious opposite of infinite Mind - called devil or evil - is not Mind, is not Truth, but error, without intelligence or reality. There can be but one Mind, because there is but one God; and if mortals claimed no other Mind and accepted no other, sin would be unknown. We can have but one Mind, if that one is infinite. We bury the sense of infinitude, when we admit that, although God is infinite, evil has a place in this infinity, for evil can have no place, where all space is filled with God.

The sole governor 469

We lose the high signification of omnipotence, when after admitting that God, or good, is omnipresent and has all-power, we still believe there is another power, named evil. This belief that there is more than one mind is as pernicious to divine theology as are ancient mythology and pagan idolatry. With one Father, even God, the whole family of man would be brethren; and with one Mind and that God, or good, the brotherhood of man would consist of Love and Truth, and have unity of Principle and spiritual power which constitute divine Science. The supposed existence of more than one mind was the basic error of idolatry. This error assumed the loss of spiritual power, the loss of the spiritual presence of Life as infinite Truth without an unlikeness, and the loss of Love as ever present and universal.

The divine standard of perfection 470

Divine Science explains the abstract statement that there is one Mind by the following self-evident proposition: If God, or good, is real, then evil, the unlikeness of God, is unreal. And evil can only seem to be real by giving reality to the unreal. The children of God have but one Mind. How can good lapse into evil, when God, the Mind of man, never sins? The standard of perfection was originally God and man. Has God taken down His own standard, and has man fallen?

Indestructible relationship 470

God is the creator of man, and, the divine Principle of man remaining perfect, the divine idea or reflection, man, remains perfect. Man is the expression of God's being. If there ever was a moment when man did not express the divine perfection, then there was a moment when man did not express God, and consequently a time when Deity was unexpressed - that - is, without entity. If man has lost perfection, then he has lost his perfect Principle, the divine Mind. If man ever existed without this perfect Principle or Mind, then man's existence was a myth.

The relations of God and man, divine Principle and idea, are indestructible in Science; and Science knows no lapse from nor return to harmony, but holds the divine order or spiritual law, in which God and all that He creates are perfect and eternal, to have remained unchanged in its eternal history.

Celestial evidence 471

The unlikeness of Truth, - named **error**, - the opposite of Science, and the evidence before the five corporeal senses, afford no indication of the grand facts of being; even as these so-called senses receive no intimation of the earth's motions or of the science of astronomy, but yield assent to astronomical propositions on the authority of natural science.

The facts of divine Science should be admitted, - although the evidence as to these facts is not supported by evil, by matter, or by material sense, - because the evidence that God and man coexist is fully sustained by spiritual sense. Man is, and forever has been, God's reflection. God is infinite, therefore ever present, and there is no other power nor presence. Hence the spirituality of the universe is the only fact of creation. "Let God be true, but every [material] man a liar."

---

oo3o

~ Question 11. - **Are doctrines and creeds a benefit to man?~**
~Adam: The first god of mythology.~
~Lesson topic: SUBSTANCE~

Answer. -
(The test of experience)
The author subscribed to an orthodox creed in early youth, and tried to adhere to it until she caught the first gleam of that which interprets God as above mortal sense. This view rebuked human beliefs, and gave the spiritual import, expressed through Science, of all that proceeds from the divine Mind. Since then her highest creed has been divine Science, which, reduced to human apprehension, she has named Christian Science. This Science teaches man that God is the only Life, and that this Life is Truth and Love; that God is to be understood, adored, and demonstrated; that divine Truth casts out suppositional error and heals the sick.

God's law destroys evil 472

The way which leads to Christian Science is straight and narrow. God has set His signet upon Science, making it coordinate with all that is real and only with that which is harmonious and eternal.

Sickness, sin, and death, being inharmonious, do not originate in God nor belong to His government. His law, rightly understood, destroys them. Jesus furnished proofs of these statements.

~ Question 12. - **What is error?**~

~Adam: Not God's man, who represents the one God and is His own image and likeness.~
~Lesson topic: MATTER~

Answer. -

(Evanescent materiality)
Error is a supposition that pleasure and pain, that intelligence, substance, life, are existent in matter. Error is neither Mind nor one of Mind's faculties. Error is the contradiction of Truth. Error is a belief without understanding. Error is unreal because untrue. It is that which seemeth to be and is not. If error were true, its truth would be error, and we should have a self-evident absurdity-namely, **erroneous truth**. Thus we should continue to lose the standard of Truth.

o3oo

~ Question 13. - **Is there no sin?**~

~Adam: The opposite of Spirit and His creations.~
~Lesson topic: REALITY~

Answer. -

(Unrealities that seem real)
All reality is in God and His creation, harmonious and eternal. That which He creates is good, and He makes all that is made. Therefore the only reality of sin, sickness, or death is the awful fact that unrealities seem real to human, erring belief, until God strips off their disguise. They are not true, because they are not of God. We learn in Christian Science that all inharmony of mortal mind or body is illusion, possessing neither reality nor identity though seeming to be real and identical.

Christ the ideal Truth 473

The Science of Mind disposes of all evil. Truth, God, is not the father of error. Sin, sickness, and death are to be classified as effects of error. Christ came to destroy the belief of sin. The God-principle is omnipresent and omnipotent. God is everywhere, and nothing apart from Him is present or has power. Christ is the ideal Truth, that comes to heal sickness and sin through Christian Science, and attributes all power to God. Jesus is the name of the man who, more than all other men, has presented Christ, the true idea of God, healing the sick and the sinning and destroying the power of death. Jesus is the human man, and Christ is the divine idea; hence the duality of Jesus the Christ.

Jesus not God 473

In an age of ecclesiastical despotism, Jesus introduced the teaching and practice of Christianity, affording the proof of Christianity's truth and love; but to reach his example and to test its unerring Science according to his rule, healing sickness, sin, and death, a better understanding of God as divine Principle, Love, rather than personality or the man Jesus, is required.

Jesus not understood 473

Jesus established what he said by demonstration, thus making his acts of higher importance than his words. He proved what he taught. This is the Science of Christianity. Jesus proved the Principle, which heals the sick and casts out error, to be divine. Few, however, except his students understood in the least his teachings and their glorious proofs, - namely, that Life, Truth, and Love (the Principle of this unacknowledged Science) destroy all error, evil, disease, and death.

Miracles rejected 474

The reception accorded to Truth in the early Christian era is repeated to-day. Whoever introduces the Science of Christianity will be scoffed at and scourged with worse cords than those which cut the flesh. To the ignorant age in which it first appears, Science seems to be a mistake, - hence the misinterpretation and consequent maltreatment which it receives. Christian marvels (and marvel is the simple meaning of the Greek word rendered miracle in the New Testament) will be misunderstood and misused by many, until the glorious Principle of these marvels is gained.

Divine fulfillment 474

If sin, sickness, and death are as real as Life, Truth, and Love, then they must all be from the same source; God must be their author. Now Jesus came to destroy sin, sickness, and death; yet the Scriptures aver, "I am not come to destroy, but to fulfil." Is it possible, then, to believe that the evils which Jesus lived to destroy are real or the offspring of the divine will?

Truth destroys falsity 474

Despite the hallowing influence of Truth in the destruction of error, must error still be immortal? Truth spares all that is true. If evil is real, Truth must make it so; but error, not Truth, is the author of the unreal, and the unreal vanishes, while all that is real is eternal. The apostle says that the mission of Christ is to "destroy the works of the devil." Truth destroys falsity and error, for light and darkness cannot dwell together. Light extinguishes the darkness, and the Scripture declares that there is "no night there." To Truth there is no error, - all is Truth. To infinite Spirit there is no matter, - all is Spirit, divine Principle and its idea.

~ Question 14. - **What is man?**~

~Adam: That which is not the image and likeness of good, but a material belief, opposed to the one Mind, or Spirit.~

~Lesson topic: UNREALITY~

Answer. -

(Fleshly factors unreal)

Man is not matter; he is not made up of brain, blood, bones, and other material elements. The Scriptures inform us that man is made in the image and likeness of God. Matter is not that likeness. The likeness of Spirit cannot be so unlike Spirit. Man is spiritual and perfect; and because he is spiritual and perfect, he must be so understood in Christian Science. Man is idea, the image, of Love; he is not physique. He is the compound idea of God, including all right ideas; the generic term for all that reflects God's image and likeness; the conscious identity of being as found in Science, in which man is the reflection of God, or Mind, and therefore is eternal; that which has no separate mind from God; that which has not a single quality underived from Deity; that which possesses no life, intelligence, nor creative power of his own, but reflects spiritually all that belongs to his Maker.

And God said: "Let us make man in our image, after our likeness; and let them have dominion over the fish of the sea, and over the fowl of the air, and over the cattle, and over all the earth, and over every creeping thing that creepeth upon the earth."

Man unfallen 475

Man is incapable of sin, sickness, and death. The real man cannot depart from holiness, nor can God, by whom man is evolved, engender the capacity or freedom to sin. A mortal sinner is not God's man. Mortals are the counterfeits of immortals. They are the children of the wicked one, or the one evil, which declares that man begins in dust or as a material embryo. In divine Science, God and the real man are inseparable as divine Principle and idea.

Mortals are not immortals 476

Error, urged to its final limits, is self-destroyed. Error will cease to claim that soul is in body, that life and intelligence are in matter, and that this matter is man. God is the Principle of man, and man is the idea of God. Hence man is not mortal nor material. Mortals will disappear, and immortals, or the children of God, will appear as the only and eternal verities of man. Mortals are not fallen children of God. They never had a perfect state of being, which may subsequently be regained. They were, from the beginning of mortal history, "conceived in sin and brought forth in iniquity." Mortality is finally swallowed up in immortality. Sin, sickness, and death must disappear to give place to the facts which belong to immortal man.

Imperishable identity 476

Learn this, O mortal, and earnestly seek the spiritual status of man, which is outside of all material selfhood. Remember that the Scriptures say of mortal man: "As for man, his days are as grass: as a flower of the field, so he flourisheth. For the wind passeth over it, and it is gone; and the place thereof shall know it no more."

The kingdom within 476

When speaking of God's children, not the children of men, Jesus said, "The kingdom of God is within you;" that is, Truth and Love reign in the real man, showing that man in God's image is unfallen and eternal. Jesus beheld in Science the perfect man, who appeared to him where sinning mortal man appears to mortals. In this perfect man the Saviour saw God's own likeness, and this correct view of man healed the sick. Thus Jesus taught that the kingdom of God is intact, universal, and that man is pure and holy. Man is not a material habitation for Soul; he is himself spiritual. Soul, being Spirit, is seen in nothing imperfect nor material.

Material body never God's idea 477

Whatever is material is mortal. To the five corporeal senses, man appears to be matter and mind united; but Christian Science reveals man as the idea of God, and declares the corporeal senses to be mortal and erring illusions. Divine Science shows it to be impossible that a material body, though interwoven with matter's highest stratum, misnamed mind, should be man, - the genuine and perfect man, the immortal idea of being, indestructible and eternal. Were it otherwise, man would be annihilated.

3ooo

~ Question 15. - **What are body and Soul?**~
~Adam: A co-called finite mind, producing other minds, thus making "gods many and lords many" - I Corinthians viii. 5~
~Lesson topic: ARE SIN, DISEASE, AND DEATH REAL?~

Answer. -
(Reflection of Spirit)
Identity is the reflection of Spirit, the reflection in multifarious forms of the living Principle, Love. Soul is the substance, Life, and intelligence of man, which is individualized, but not in matter. Soul can never reflect anything inferior to Spirit.

Man inseparable from Spirit 477

Man is the expression of Soul. The Indians caught some glimpses of the underlying reality, when they called a certain beautiful lake "the smile of the Great Spirit." Separated from man, who expresses Soul, Spirit would be a nonentity; man, divorced from Spirit, would lose his entity. But there is, there can be, no such division, for man is coexistent with God.

A vacant domicile 478

What evidence of Soul or of immortality have you within mortality? Even according to the teachings of natural science, man has never beheld Spirit or Soul leaving a body or entering it. What basis is there for the theory of indwelling spirit, except the claim of mortal belief? What would be thought of the declaration that a house was inhabited, and by a certain class of persons, when no such persons were ever seen to go into the house or to come out of it, nor were they even visible through the windows? Who can see a soul in the body?

~ Question 16. - Does brain think, and do nerves feel, and is there intelligence in matter?~

~Adam: A product of nothing as the mimicry of something.~
~Lesson topic: DOCTRINE OF ATONEMENT~

Answer. -

(Harmonious functions)
No, not if God is true and mortal man a liar. The assertion that there can be pain or pleasure in matter is erroneous. That body is most harmonious in which the discharge of the natural functions is least noticeable. How can intelligence dwell in matter when matter is non-intelligent and brain-lobes cannot think? Matter cannot perform the functions of Mind. Error says, "I am man;" but this belief is mortal and far from actual. From beginning to end, whatever is mortal is composed of material human beliefs and of nothing else. That only is real which reflects God. St. Paul said, "But when it pleased God, who separated me from my mother's womb, and called me by His grace, . . . I conferred not with flesh and blood."

Immortal birthright 478

Mortal man is really a self-contradictory phrase, for man is not mortal, "neither indeed can be;" man is immortal. If a child is the offspring of physical sense and not of Soul, the child must have a material, not a spiritual origin. With what truth, then, could the Scriptural rejoicing be uttered by any mother, "I have gotten a man from the Lord"? On the contrary, if aught comes from God, it cannot be mortal and material; it must be immortal and spiritual.

Matter's supposed selfhood 479

Matter is neither self-existent nor a product of Spirit. An image of mortal thought, reflected on the retina, is all that the eye beholds. Matter cannot see, feel, hear, taste, nor smell. It is not self-cognizant, - cannot feel itself, see itself, nor understand itself. Take away so-called mortal mind, which constitutes matter's supposed selfhood, and matter can take no cognizance of matter. Does that which we call dead ever see, hear, feel, or use any of the physical senses?

Chaos and darkness 479

"In the beginning God created the heaven and the earth. And the earth was without form, and void; and darkness was upon the face of the deep." (Genesis i. 1, 2.) In the vast forever, in the Science and truth of being, the only facts are Spirit and its innumerable creations. Darkness and chaos are the imaginary opposites of light, understanding, and eternal harmony, and they are the elements of nothingness.

Spiritual reflection 479

We admit that black is not a color, because it reflects no light. So evil should be denied identity or power, because it has none of the divine hues. Paul says: "For the invisible things of Him, from the creation of the world, are clearly seen, being understood by the things that are made." (Romans i. 20.) When the substance of Spirit appears in Christian Science, the nothingness of matter is recognized. Where the spirit of God is, and there is no place where God is not, evil becomes nothing, - the opposite of the something of Spirit. If there is no spiritual reflection, then there remains only the darkness of vacuity and not a trace of heavenly tints.

Harmony from Spirit 480

Nerves are an element of the belief that there is sensation in matter, whereas matter is devoid of sensation. Consciousness, as well as action, is governed by Mind, - is in God, the origin and governor of all that Science reveals. Material sense has its realm apart from Science in the unreal. Harmonious action proceeds from Spirit, God. Inharmony has no Principle; its action is erroneous and presupposes man to be in matter. Inharmony would make matter the cause as well as the effect of intelligence, or Soul, thus attempting to separate Mind from God.

Evil nonexistent 480

Man is not God, and God is not man. Again, God, or good, never made man capable of sin. It is the opposite of good - that is, evil - which seems to make men capable of wrong-doing. Hence, evil is but an illusion, and it has no real basis. Evil is a false belief. God is not its author. The supposititious parent of evil is a lie.

Vapor and nothingness 480

The Bible declares: "All things were made by Him the divine Word]; and without Him was not anything made that was made." This is the eternal verity of divine Science. If sin, sickness, and death were understood as nothingness, they would disappear. As vapor melts before the sun, so evil would vanish before the reality of good. One must hide the other. How important, then, to choose good as the reality! Man is tributary to God, Spirit, and to nothing else. God's being is infinity, freedom, harmony, and boundless bliss. "Where the Spirit of the Lord is, there is liberty." Like the archpriests of yore, man is free "to enter into the holiest," - the realm of God.

The fruit forbidden 481

Material sense never helps mortals to understand Spirit, God. Through spiritual sense only, man comprehends and loves Deity. The various contradictions of the Science of Mind by the material senses do not change the unseen Truth, which remains forever intact. The forbidden fruit of knowledge, against which wisdom warns man, is the testimony of error, declaring existence to be at the mercy of death, and good and evil to be capable of commingling. This is the significance of the Scripture concerning this "tree of the knowledge of good and evil," - this growth of material belief, of which it is said: "In the day that thou eatest thereof thou shalt surely die." Human hypotheses first assume the reality of sickness, sin, and death, and then assume the necessity of these evils because of their admitted actuality. These human verdicts are the procurers of all discord.

Sense and pure Soul 481

If Soul sins, it must be mortal. Sin has the elements of self-destruction. It cannot sustain itself. If sin is supported, God must uphold it, and this is impossible, since Truth cannot support error. Soul is the divine Principle of man and never sins, - hence the immortality of Soul. In Science we learn that it is material sense, not Soul, which sins; and it will be found that it is the sense of sin which is lost, and not a sinful soul. When reading the Scriptures, the substitution of the word sense for soul gives the exact meaning in a majority of cases.

Soul defined 482

Human thought has adulterated the meaning of the word soul through the hypothesis that soul is both an evil and a good intelligence, resident in matter. The proper use of the word soul can always be gained by substituting the word God, where the deific meaning is required. In other cases, use the word sense, and you will have the scientific signification. As used in Christian Science, Soul is properly the synonym of Spirit, or God; but out of Science, soul is identical with sense, with material sensation.

0004

~ Question 17. - **Is it important to understand these explanations**

in order to heal the sick?~

~Adam: An unreality as opposed to the great reality of spiritual existence and creation.~
~Lesson topic: PROBATION AFTER DEATH~

Answer. -

(Sonship of Jesus)

It is, since Christ is "the way" and the truth casting out all error. Jesus called himself "the Son of man," but not the son of Joseph. As woman is but a species of the genera, he was literally the Son of Man. Jesus was the highest human concept of the perfect man. He was inseparable from Christ, the Messiah, - the divine idea of God outside the flesh. This enabled Jesus to demonstrate his control over matter. Angels announced to the Wisemen of old this dual appearing, and angels whisper it, through faith, to the hungering heart in every age.

Sickness erroneous 482

Sickness is part of the error which Truth casts out. Error will not expel error. Christian Science is the law of Truth, which heals the sick on the basis of the one Mind or God. It can heal in no other way, since the human, mortal mind so-called is not a healer, but causes the belief in disease.

Then comes the question, how do drugs, hygiene, and animal magnetism heal? It may be affirmed that they do not heal, but only relieve suffering temporarily, exchanging one disease for another. We classify disease as error, which nothing but Truth or Mind can heal, and this Mind must be divine, not human. Mind transcends all other power, and will ultimately supersede all other means in healing. In order to heal by Science, you must not be ignorant of the moral and spiritual demands of Science nor disobey them. Moral ignorance or sin affects your demonstration, and hinders its approach to the standard in Christian Science.

Terms adopted by the author 483

After the author's sacred discovery, she affixed the name "Science" to Christianity, the name "error" to corporeal sense, and the name "substance" to Mind. Science has called the world to battle over this issue and its demonstration, which heals the sick, destroys error, and reveals the universal harmony. To those natural Christian Scientists, the ancient worthies, and to Christ Jesus, God certainly revealed the spirit of Christian Science, if not the absolute letter.

Science the way 483

Because the Science of Mind seems to bring into dishonor the ordinary scientific schools, which wrestle with material observations alone, this Science has met with opposition; but if any system honors God, it ought to receive aid, not opposition, from all thinking persons. And Christian Science does honor God as no other theory honors Him, and it does this in the way of His appointing, by doing many wonderful works through the divine name and nature. One must fulfil one's mission without timidity or dissimulation, for to be well done, the work must be done unselfishly. Christianity will never be based on a divine Principle and so found to be unerring, until its absolute Science is reached.

When this is accomplished, neither pride, prejudice, bigotry, nor envy can wash away its foundation, for it is built upon the rock, Christ.

~ Question 18. - Does Christian Science, or metaphysical healing, include medication, material hygiene, mesmerism, hypnotism, theosophy, or spiritualism?~

~Adam: A so-called man, whose origin, substance, and mind are found to be the antipode of God, or Spirit.~

~Lesson topic: EVERLASTING PUNISHMENT~

Answer. -

(Mindless methods)

Not one of them is included in it. In divine Science, the supposed laws of matter yield to the law of Mind. What are termed natural science and material laws are the objective states of mortal mind. The physical universe expresses the conscious and unconscious thoughts of mortals. Physical force and mortal mind are one. Drugs and hygiene oppose the supremacy of the divine Mind. Drugs and inert matter are unconscious, mindless. Certain results, supposed to proceed from drugs, are really caused by the faith in them which the false human consciousness is educated to feel.

Mesmerism is mortal, material illusion. Animal magnetism is the voluntary or involuntary action of error in all its forms; it is the human antipode of divine Science. Science must triumph over material sense, and Truth over error, thus putting an end to the hypotheses involved in all false theories and practices.

---

oo4o

~ Question 19. - Is materiality the concomitant of spirituality, and is material sense a necessary preliminary to the understanding and expression of Spirit?~

~Adam: An inverted image of Spirit.~

~Lesson topic: ADAM AND FALLEN MAN~

Answer. -

(Error only ephemeral)

If error is necessary to define or to reveal Truth, the answer is yes; but not otherwise. Material sense is an absurd phrase, for matter has no sensation. Science declares that Mind, not matter, sees, hears, feels, speaks. Whatever contradicts this statement is the false sense, which ever betrays mortals into sickness, sin, and death. If the unimportant and evil appear, only soon to disappear because of their uselessness or their iniquity, then these ephemeral views of error ought to be obliterated by Truth. Why malign Christian Science for instructing mortals how to make sin, disease, and death appear more and more unreal?

Scientific translations 485

Emerge gently from matter into Spirit. Think not to thwart the spiritual ultimate of all things, but come naturally into Spirit through better health and morals and as the result of spiritual growth. Not death, but the understanding of Life, makes man immortal. The belief that life can be in matter or soul in body, and that man springs from dust or from an egg, is the result of the mortal error which Christ, or Truth, destroys by fulfilling the spiritual law of being, in which man is perfect, even as the "Father which is in heaven is perfect." If thought yields its dominion to other powers, it cannot outline on the body its own beautiful images, but it effaces them and delineates foreign agents, called disease and sin.

Material beliefs 485

The heathen gods of mythology controlled war and agriculture as much as nerves control sensation or muscles measure strength. To say that strength is in matter, is like saying that the power is in the lever. The notion of any life or intelligence in matter is without foundation in fact, and you can have no faith in falsehood when you have learned falsehood's true nature.

Sense 'versus' Soul 486

Suppose one accident happens to the eye, another to the ear, and so on, until every corporeal sense is quenched. What is man's remedy? To die, that he may regain these senses? Even then he must gain spiritual understanding and spiritual sense in order to possess immortal consciousness. Earth's preparatory school must be improved to the utmost. In reality man never dies. The belief that he dies will not establish his scientific harmony. Death is not the result of Truth but of error, and one error will not correct another.

Death an error 486

Jesus proved by the prints of the nails, that his body was the same immediately after death as before. If death restores sight, sound, and strength to man, then death is not an enemy but a better friend than Life. Alas for the blindness of belief, which makes harmony conditional upon death and matter, and yet supposes Mind unable to produce harmony! So long as this error of belief remains, mortals will continue mortal in belief and subject to chance and change.

Permanent sensibility 486

Sight, hearing, all the spiritual senses of man, are eternal. They cannot be lost. Their reality and immortality are in Spirit and understanding, not in matter, - hence their permanence. If this were not so, man would be speedily annihilated. If the five corporeal senses were the medium through which to understand God, then palsy, blindness, and deafness would place man in a terrible situation, where he would be like those "having no hope, and without God in the world;" but as a matter of fact, these calamities often drive mortals to seek and to find a higher sense of happiness and existence.

Exercise of Mind faculties 487

Life is deathless. Life is the origin and ultimate of man, never attainable through death, but gained by walking in the pathway of Truth both before and after that which is called death. There is more Christianity in seeing and hearing spiritually than materially. There is more Science in the perpetual exercise of the Mind-faculties than in their loss. Lost they cannot be, while Mind remains. The apprehension of this gave sight to the blind and hearing to the deaf centuries ago, and it will repeat the wonder.

~ Question 20. - You speak of belief. Who or what is it that believes?~

~Adam: The image and likeness of what God has not created, namely, matter, sin, sickness, and death.~
~Lesson topic: MORTALS AND IMMORTALS~

~Adam: The opposer of Truth, termed error.~
~Lesson topic: SOUL AND BODY~

Answer. -

(Understanding versus belief)
Spirit is all-knowing; this precludes the need of believing. Matter cannot believe, and Mind understands. The body cannot believe. The believer and belief are one and are mortal. Christian evidence is founded on Science or demonstrable Truth, flowing from immortal Mind, and there is in reality no such thing as mortal mind. Mere belief is blindness without Principle from which to explain the reason of its hope. The belief that life is sentient and intelligent matter is erroneous.

The Apostle James said, "Show me thy faith without thy works, and I will show thee my faith by my works." The understanding that Life is God, Spirit, lengthens our days by strengthening our trust in the deathless reality of Life, its almightiness and immortality.

Confirmation by healing 487

This faith relies upon an understood Principle. This Principle makes whole the diseased, and brings out the enduring and harmonious phases of things. The result of our teachings is their sufficient confirmation. When, on the strength of these instructions, you are able to banish a severe malady, the cure shows that you understand this teaching, and therefore you receive the blessing of Truth.

Belief and firm trust 488

The Hebrew and Greek words often translated belief differ somewhat in meaning from that conveyed by the English verb believe; they have more the significance of faith, understanding, trust, constancy, firmness. Hence the Scriptures often appear in our common version to approve and endorse belief, when they mean to enforce the necessity of understanding.

---

o4oo

~ Question 21. - Do the five corporeal senses constitute man?~
~Adam: Life's counterfeit, which ultimates in death.~
~Lesson topic: ANCIENT AND MODERN NECROMANY, ALIAS MESMERISM AND HYPNOTISM DENOUNCED~

Answer. -
(All faculties from Mind)
Christian Science sustains with immortal proof the impossibility of any material sense, and defines these so-called senses as mortal beliefs, the testimony of which cannot be true either of man or of his Maker. The corporeal senses can take no cognizance of spiritual reality and immortality. Nerves have no more sensation, apart from what belief bestows upon them, than the fibres of a plant. Mind alone possesses all faculties, perception, and comprehension. Therefore mental endowments are not at the mercy of organization and decomposition, - otherwise the very worms could unfashion man. If it were possible for the real senses of man to be injured, Soul could reproduce them in all their perfection; but they cannot be disturbed nor destroyed, since they exist in immortal Mind, not in matter.

Possibilities of Life 489

The less mind there is manifested in matter the better. When the unthinking lobster loses its claw, the claw grows again. If the Science of Life were understood, it would be found that the senses of Mind are never lost and that matter has no sensation. Then the human limb would be replaced as readily as the lobster's claw, - not with an artificial limb, but with the genuine one. Any hypothesis which supposes life to be in matter is an educated belief. In infancy this belief is not equal to guiding the hand to the mouth; and as consciousness develops, this belief goes out, - yields to the reality of everlasting Life.

Decalogue disregarded 489

Corporeal sense defrauds and lies; it breaks all the commands of the Mosaic Decalogue to meet its own demands. How then can this sense be the God-given channel to man of divine blessings or understanding? How can man, reflecting God, be dependent on material means for knowing, hearing, seeing? Who dares to say that the senses of man can be at one time the medium for sinning against God, at another the medium for obeying God? An affirmative reply would contradict the Scripture, for the same fountain sendeth not forth sweet waters and bitter.

Organic construction valueless 489

The corporeal senses are the only source of evil or error. Christian Science shows them to be false, because matter has no sensation, and no organic construction can give it hearing and sight nor make it the medium of Mind. Outside the material sense of things, all is harmony. A wrong sense of God, man, and creation is non-sense, want of sense. Mortal belief would have the material senses sometimes good and sometimes bad. It assures mortals that there is real pleasure in sin; but the grand truths of Christian Science dispute this error.

Will-power an animal propensity 490

Will-power is but a product of belief, and this belief commits depredations on harmony. Human will is an animal propensity, not a faculty of Soul. Hence it cannot govern man aright. Christian Science reveals Truth and Love as the motive-powers of man. Will-blind, stubborn, and headlong - cooperates with appetite and passion. From this cooperation arises its evil. From this also comes its powerlessness, since all power belongs to God, good.

Theories helpless 490

The Science of Mind needs to be understood. Until it is understood, mortals are more or less deprived of Truth. Human theories are helpless to make man harmonious or immortal, since he is so already, according to Christian Science. Our only need is to know this and reduce to practice the real man's divine Principle, Love.

True nature and origin 490

"Quench not the Spirit. Despise not prophesyings." Human belief - or knowledge gained from the so-called material senses - would, by fair logic, annihilate man along with the dissolving elements of clay. The scientifically Christian explanations of the nature and origin of man destroy all material sense with immortal testimony. This immortal testimony ushers in the spiritual sense of being, which can be obtained in no other way.

Sleep an illusion 490

Sleep and mesmerism explain the mythical nature of material sense. Sleep shows material sense as either oblivion, nothingness, or an illusion or dream. Under the mesmeric illusion of belief, a man will think that he is freezing when he is warm, and that he is swimming when he is on dry land. Needle-thrusts will not hurt him. A delicious perfume will seem intolerable. Animal magnetism thus uncovers material sense, and shows it to be a belief without actual foundation or validity. Change the belief, and the sensation changes. Destroy the belief, and the sensation disappears.

Material man is made up of involuntary and voluntary error, of a negative right and a positive wrong, the latter calling itself right. Man's spiritual individuality is never wrong. It is the likeness of man's Maker. Matter cannot connect mortals with the true origin and facts of being, in which all must end. It is only by acknowledging the supremacy of Spirit, which annuls the claims of matter, that mortals can lay off mortality and find the indissoluble spiritual link which establishes man forever in the divine likeness, inseparable from his creator.

Material man as a dream 491

The belief that matter and mind are one, - that matter is awake at one time and asleep at another, sometimes presenting no appearance of mind, - this belief culminates in another belief, that man dies. Science reveals material man as never the real being. The dream or belief goes on, whether our eyes are closed or open. In sleep, memory and consciousness are lost from the body, and they wander whither they will apparently with their own separate embodiment. Personality is not the individuality of man. A wicked man may have an attractive personality.

Spiritual existence the one fact 491

When we are awake, we dream of the pains and pleasures of matter. Who will say, even though he does not understand Christian Science, that this dream - rather than the dreamer - may not be mortal man? Who can rationally say otherwise, when the dream leaves mortal man intact in body and thought, although the so-called dreamer is unconscious? For right reasoning there should be but one fact before the thought, namely, spiritual existence. In reality there is no other existence, since Life cannot be united to its unlikeness, mortality.

Being is holiness, harmony, immortality. It is already proved that a knowledge of this, even in small degree, will uplift the physical and moral standard of mortals, will increase longevity, will purify and elevate character. Thus progress will finally destroy all error, and bring immortality to light. We know that a statement proved to be good must be correct. New thoughts are constantly obtaining the floor. These two contradictory theories - that matter is something, or that all is Mind - will dispute the ground, until one is acknowledged to be the victor. Discussing his campaign, General Grant said: "I propose to fight it out on this line, if it takes all summer." Science says: All is Mind and Mind's idea. You must fight it out on this line. Matter can afford you no aid.

Scientific ultimatum 492

The notion that mind and matter commingle in the human illusion as to sin, sickness, and death must eventually submit to the Science of Mind, which denies this notion. God is Mind, and God is infinite; hence all is Mind. On this statement rests the Science of being, and the Principle of this Science is divine, demonstrating harmony and immortality.

Victory for Truth 492

The conservative theory, long believed, is that there are two factors, matter and mind, uniting on some impossible basis. This theory would keep truth and error always at war. Victory would perch on neither banner. On the other hand, Christian Science speedily shows Truth to be triumphant. To corporeal sense, the sun appears to rise and set, and the earth to stand still; but astronomical science contradicts this, and explains the solar system as working on a different plan. All the evidence of physical sense and all the knowledge obtained from physical sense must yield to Science, to the immortal truth of all things.

~ Question 22. - Will you explain sickness and show how it is to be healed?~

~Adam: The opposite of Love, called hate.~
~Lesson topic: GOD THE ONLY CAUSE AND CREATOR~

~Adam: The userper of Spirit's creation, called self-creative matter.~
~Lesson topic: GOD THE PRESERVER OF MAN~

Answer. -
(Mental preparation)
The method of Christian Science Mind-healing is touched upon in a previous chapter entitled Christian Science Practice. A full answer to the above question involves teaching, which enables the healer to demonstrate and prove for himself the Principle and rule of Christian Science or metaphysical healing.

Mind destroys all ills 493

Mind must be found superior to all the beliefs of the five corporeal senses, and able to destroy all ills. Sickness is a belief, which must be annihilated by the divine Mind. Disease is an experience of so-called mortal mind. It is fear made manifest on the body. Christian Science takes away this physical sense of discord, just as it removes any other sense of moral or mental inharmony. That man is material, and that matter suffers, - these propositions can only seem real and natural in illusion. Any sense of soul in matter is not the reality of being.

If Jesus awakened Lazarus from the dream, illusion, of death, this proved that the Christ could improve on a false sense. Who dares to doubt this consummate test of the power and willingness of divine Mind to hold man forever intact in his perfect state, and to govern man's entire action? Jesus said: "Destroy this temple [body], and in three days I [Mind] will raise it up;" and he did this for tired humanity's reassurance.

Inexhaustible divine Love 494

Is it not a species of infidelity to believe that so great a work as the Messiah's was done for himself or for God, who needed no help from Jesus' example to preserve the eternal harmony? But mortals did need this help, and Jesus pointed the way for them. Divine Love always has met and always will meet every human need. It is not well to imagine that Jesus demonstrated the divine power to heal only for a select number or for a limited period of time, since to all mankind and in every hour, divine Love supplies all good.

Reason and Science 494

The miracle of grace is no miracle to Love. Jesus demonstrated the inability of corporeality, as well as the infinite ability of Spirit, thus helping erring human sense to flee from its own convictions and seek safety in divine Science. Reason, rightly directed, serves to correct the errors of corporeal sense; but sin, sickness, and death will seem real (even as the experiences of the sleeping dream seem real) until the Science of man's eternal harmony breaks their illusion with the unbroken reality of scientific being.

Which of these two theories concerning man are you ready to accept? One is the mortal testimony, changing, dying, unreal. The other is the eternal and real evidence, bearing Truth's signet, its lap piled high with immortal fruits.

Our Master cast out devils (evils) and healed the sick. It should be said of his followers also, that they cast fear and all evil out of themselves and others and heal the sick. God will heal the sick through man, whenever man is governed by God. Truth casts out error now as surely as it did nineteen centuries ago. All of Truth is not understood; hence its healing power is not fully demonstrated.

Destruction of all evil 495

If sickness is true or the idea of Truth, you cannot destroy sickness, and it would be absurd to try. Then classify sickness and error as our Master did, when he spoke of the sick, "whom Satan hath bound," and find a sovereign antidote for error in the life-giving power of Truth acting on human belief, a power which opens the prison doors to such as are bound, and sets the captive free physically and morally.

Steadfast and calm trust 495

When the illusion of sickness or sin tempts you, cling steadfastly to God and His idea. Allow nothing but His likeness to abide in your thought. Let neither fear nor doubt overshadow your clear sense and calm trust, that the recognition of life harmonious - as Life eternally is - can destroy any painful sense of, or belief in, that which Life is not. Let Christian Science, instead of corporeal sense, support your understanding of being, and this understanding will supplant error with Truth, replace mortality with immortality, and silence discord with harmony.

4ooo

~ Question 23. - How can I progress most rapidly in the understanding of Christian Science?~

~Adam: Immortality's opposite, mortality.~
~Lesson topic: IS THE UNIVERSE INCLUSDING MAN EVOLVED BY ATOMIC FORCE?~

Answer. -

(Rudiments of growth)
Study thoroughly the letter and imbibe the spirit. Adhere to the divine Principle of Christian Science and follow the behests of God, abiding steadfastly in wisdom, Truth, and Love. In the Science of Mind, you will soon ascertain that error cannot destroy error. You will also learn that in Science there is no transfer of evil suggestions from one mortal to another, for there is but one Mind, and this ever-present omnipotent Mind is reflected by man and governs the entire universe. You will learn that in Christian Science the first duty is to obey God, to have one Mind, and to love another as yourself.

We all must learn that Life is God. Ask yourself: Am I living the life that approaches the supreme good? Am I demonstrating the healing power of Truth and Love? If so, then the way will grow brighter "unto the perfect day." Your fruits will prove what the understanding of God brings to man. Hold perpetually this thought, - that it is the spiritual idea, the Holy Ghost and Christ, which enables you to demonstrate, with scientific certainty, the rule of healing, based upon its divine Principle, Love, underlying, overlying, and encompassing all true being.

Triumph over death 496

"The sting of death is sin; and the strength of sin is the law," - the law of mortal belief, at war with the facts of immortal Life, even with the spiritual law which says to the grave, "Where is thy victory?" But "when this corruptible shall have put on incorruption, and this mortal shall have put on immortality, then shall be brought to pass the saying that is written, Death is swallowed up in victory."

~ Question 24. - Have Christian Scientists any religious greed?~

~Adam: That of which wisdom saith, "Thou shalt surely die."~

~Lesson topic: CHRISTIAN SCIENCE~

Answer. - They have not, if by that term is meant doctrinal beliefs. The following is a brief exposition of the important points, or religious tenets, of Christian Science:-
(- Tenets -)

1. As adherents of Truth, we take the inspired Word of the Bible as our sufficient guide to eternal Life.

2. We acknowledge and adore one supreme and infinite God. We acknowledge His Son, one Christ; the Holy Ghost or divine Comforter; and man in God's image and likeness.

3. We acknowledge God's forgiveness of sin in the destruction of sin and the spiritual understanding that casts out evil as unreal. But the belief in sin is punished so long as the belief lasts.

4. We acknowledge Jesus' atonement as the evidence of divine, efficacious Love, unfolding man's unity with God through Christ Jesus the Way-shower; and we acknowledge that man is saved through Christ, through Truth, Life, and Love as demonstrated by the Galilean Prophet in healing the sick and overcoming sin and death.

5. We acknowledge that the crucifixion of Jesus and his resurrection served to uplift faith to understand eternal Life, even the allness of Soul, Spirit, and the nothingness of matter.

6. And we solemnly promise to watch, and pray for that Mind to be in us which was also in Christ Jesus; to do unto others as we would have them do unto us; and to be merciful, just, and pure.

The Glossary

of the Christian Science textbook, Science and Health with Key to the Scriptures, by Mary Baker Eddy

These things saith He that is holy, He that is true, He that hath the key of David, He that openeth, and no man shutteth; and shutteth, and no man openeth; I know thy works: behold, I have set before thee an open door, and no man can shut it. - Revelation.

In Christian Science we learn that the substitution of the spiritual for the material definition of a Scriptural word often elucidates the meaning of the inspired writer. On this account this chapter is added. It contains the metaphysical interpretation of Bible terms, giving their spiritual sense, which is also their original meaning.

Abel. Watchfulness; self-offering; surrendering to the creator the early fruits of experience.

Abraham. Fidelity; faith in the divine Life and in the eternal Principle of being.

This patriarch illustrated the purpose of Love to create trust in good, and showed the life-preserving power of spiritual understanding.

Adam. Error; a falsity; the belief in "original sin," sickness, and death; evil; the opposite of good, - of God and His creation; a curse; a belief in intelligent matter, finiteness, and mortality; "dust to dust;" red sandstone; nothingness; the first god of mythology; not God's man, who represents the one God and is His own image and likeness; the opposite of Spirit and His creations; that which is not the image and likeness of good, but a material belief, opposed to the one Mind, or Spirit; a so-called finite mind, producing other minds, thus making "gods many and lords many" (I Corinthians viii. 5); a product of nothing as the mimicry of something; an unreality as opposed to the great reality of spiritual existence and creation; a so-called man, whose origin, substance, and mind are found to be the antipode of God, or Spirit; an inverted image of Spirit; the image and likeness of what God has not created, namely, matter, sin, sickness, and death; the opposer of Truth, termed error; Life's counterfeit, which ultimates in death; the opposite of Love, called hate; the usurper of Spirit's creation, called self-creative matter; immortality's 'opposite, mortality; that of which wisdom saith, "Thou shalt surely die."

The name Adam represents the false supposition that Life is not eternal, but has beginning and end; that the infinite enters the finite, that intelligence passes into non-intelligence, and that Soul dwells in material sense; that immortal Mind results in matter, and matter in mortal mind; that the one God and creator entered what He created, and then disappeared in the atheism of matter.

Adversary. An adversary is one who opposes, denies, disputes, not one who constructs and sustains reality and Truth. Jesus said of the devil, "He was a murderer from the beginning, . . . he is a liar and the father of it." This view of Satan is confirmed by the name often conferred upon him in Scripture, the "adversary."

Almighty. All-power; infinity; omnipotence.

Angels. God's thoughts passing to man; spiritual intuitions, pure and perfect; the inspiration of goodness, purity, and immortality, counteracting all evil, sensuality, and mortality.

Ark. Safety; the idea, or reflection, of Truth, proved to be as immortal as its Principle; the understanding of Spirit, destroying belief in matter.

God and man coexistent and eternal; Science showing that the spiritual realities of all things are created by Him and exist forever. The ark indicates temptation overcome and followed by exaltation.

Asher (Jacob's son). Hope and faith; spiritual compensation; the ills of the flesh rebuked.

Babel. Self-destroying error; a kingdom divided against itself, which cannot stand; material knowledge.

The higher false knowledge builds on the basis of evidence obtained from the five corporeal senses, the more confusion ensues, and the more certain is the downfall of its structure.

Baptism. Purification by Spirit; submergence in Spirit.

We are "willing rather to be absent from the body, and to be present with the Lord." (II Corinthians

v. 8.)

Believing. Firmness and constancy; not a faltering nor a blind faith, but the perception of spiritual Truth. Mortal thoughts, illusion.

Benjamin (Jacob's son). A physical belief as to life, substance, and mind; human knowledge, or so-called mortal mind, devoted to matter; pride; envy; fame; illusion; a false belief; error masquerading as the possessor of life, strength, animation, and power to act.

Renewal of affections; self-offering; an improved state of mortal mind; the introduction of a more spiritual origin; a gleam of the infinite idea of the infinite Principle; a spiritual type; that which comforts, consoles, and supports.

Bride. Purity and innocence, conceiving man in the idea of God; a sense of Soul, which has spiritual bliss and enjoys but cannot suffer.

Bridegroom. Spiritual understanding; the pure consciousness that God, the divine Principle, creates man as His own spiritual idea, and that God is the only creative power.

Burial. Corporeality and physical sense put out of sight and hearing; annihilation. Submergence in Spirit; immortality brought to light.

Canaan (the son of Ham). A sensuous belief; the testimony of what is termed material sense; the error which would make man mortal and would make mortal mind a slave to the body.

Children. The spiritual thoughts and representatives of Life, Truth, and Love.

Sensual and mortal beliefs; counterfeits of creation, whose better originals are God's thoughts, not in embryo, but in maturity; material suppositions of life, substance, and intelligence, opposed to the Science of being.

Children of Israel. The representatives of Soul, not corporeal sense; the offspring of Spirit, who, having wrestled with error, sin, and sense, are governed by divine Science; some of the ideas of God beheld as men, casting out error and healing the sick; Christ's offspring.

Christ. The divine manifestation of God, which comes to the flesh to destroy incarnate error.

Church. The structure of Truth and Love; whatever rests upon and proceeds from divine Principle.

The Church is that institution, which affords proof of its utility and is found elevating the race, rousing the dormant understanding from material beliefs to the apprehension of spiritual ideas and the demonstration of divine Science, thereby casting out devils, or error, and healing the sick.

Creator. Spirit; Mind; intelligence; the animating divine Principle of all that is real and good; self-existent Life, Truth, and Love; that which is perfect and eternal; the opposite of matter and evil, which have no Principle; God, who made all that was made and could not create an atom or an element the opposite of Himself.

Dan (Jacob's son). Animal magnetism; so-called mortal mind controlling mortal mind; error, working out the designs of error; one belief preying upon another.

Day. The irradiance of Life; light, the spiritual idea of Truth and Love.

"And the evening and the morning were the first day." (Genesis i. 5.) The objects of time and sense disappear in the illumination of spiritual understanding, and Mind measures time according to the good that is unfolded. This unfolding is God's day, and "there shall be no night there."

Death. An illusion, the lie of life in matter; the unreal and untrue; the opposite of Life.

Matter has no life, hence it has no real existence. Mind is immortal. The flesh, warring against

Spirit; that which frets itself free from one belief only to be fettered by another, until every belief of life where Life is not yields to eternal Life. Any material evidence of death is false, for it contradicts the spiritual facts of being.

Devil. Evil; a lie; error; neither corporeality nor mind; the opposite of Truth; a belief in sin, sickness, and death; animal magnetism or hypnotism; the lust of the flesh, which saith: "I am life and intelligence in matter. There is more than one mind, for I am mind, - a wicked mind, self-made or created by a tribal god and put into the opposite of mind, termed matter, thence to reproduce a mortal universe, including man, not after the image and likeness of Spirit, but after its own image."

Dove. A symbol of divine Science; purity and peace; hope and faith.

Dust. Nothingness; the absence of substance, life, or intelligence.

Ears. Not organs of the so-called corporeal senses, but spiritual understanding.

Jesus said, referring to spiritual perception, "Having ears, hear ye not?" (Mark viii. 18.)

Earth. A sphere; a type of eternity and immortality, which are likewise without beginning or end.

To material sense, earth is matter; to spiritual sense, it is a compound idea.

Elias. Prophecy; spiritual evidence opposed to material sense; Christian Science, with which can be discerned the spiritual fact of whatever the material senses behold; the basis of immortality.

"Elias truly shall first come and restore all things." (Matthew xvii. 11.)

Error. See chapter on Recapitulation, page 472.

Euphrates (river). Divine Science encompassing the universe and man; the true idea of God; a type of the glory which is to come; metaphysics taking the place of physics; the reign of righteousness. The atmosphere of human belief before it accepts sin, sickness, or death; a state of mortal thought, the only error of which is limitation; finity; the opposite of infinity.

Eve. A beginning; mortality; that which does not last forever; a finite belief concerning life, substance, and intelligence in matter; error; the belief that the human race originated materially instead of spiritually, - that man started first from dust, second from a rib, and third from an egg.

Evening. Mistiness of mortal thought; weariness of mortal mind; obscured views; peace and rest.

Eyes. Spiritual discernment, - not material but mental.

Jesus said, thinking of the outward vision, "Having eyes, see ye not?" (Mark viii. 18.)

Fan. Separator of fable from fact; that which gives action to thought.

Father. Eternal Life; the one Mind; the divine Principle, commonly called God.

Fear. Heat; inflammation; anxiety; ignorance; error; desire; caution.

Fire. Fear; remorse; lust; hatred; destruction; affliction purifying and elevating man.

Firmament. Spiritual understanding; the scientific line of demarcation between Truth and error, between Spirit and so-called matter.

Flesh. An error of physical belief; a supposition that life, substance, and intelligence are in matter; an illusion; a belief that matter has sensation.

Gad (Jacob's son). Science; spiritual being understood; haste towards harmony.

Gethsemane. Patient woe; the human yielding to the divine; love meeting no response, but still remaining love.

Ghost. An illusion; a belief that mind is outlined and limited; a supposition that spirit is finite.

Gihon (river). The rights of woman acknowledged morally, civilly, and socially.

God. The great I AM; the all-knowing, all-seeing, all-acting, all-wise, all-loving, and eternal; Principle; Mind; Soul; Spirit; Life; Truth; Love; all substance; intelligence.

Gods. Mythology; a belief that life, substance, and intelligence are both mental and material; a supposition of sentient physicality; the belief that infinite Mind is in finite forms; the various theories that hold mind to be a material sense, existing in brain, nerve, matter; supposititious minds, or souls, going in and out of matter, erring and mortal; the serpents of error, which say, "Ye shall be as gods."

God is one God, infinite and perfect, and cannot become finite and imperfect.

Good. God; Spirit; omnipotence; omniscience; omnipresence; omni-action.

Ham (Noah's son). Corporeal belief; sensuality; slavery; tyranny.

Heart. Mortal feelings, motives, affections, joys, and sorrows.

Heaven. Harmony; the reign of Spirit; government by divine Principle; spirituality; bliss; the atmosphere of Soul.

Hell. Mortal belief; error; lust; remorse; hatred; revenge; sin; sickness; death; suffering and self-destruction; self-imposed agony; effects of sin; that which "worketh abomination or maketh a lie."

Hiddekel (river). Divine Science understood and acknowledged.

Holy Ghost. Divine Science; the development of eternal Life, Truth, and Love.

I, or Ego. Divine Principle; Spirit; Soul; incorporeal, unerring, immortal, and eternal Mind.

There is but one I, or Us, but one divine Principle, or Mind, governing all existence; man and woman unchanged forever in their individual characters, even as numbers which never blend with each other, though they are governed by one Principle. All the objects of God's creation reflect one Mind, and whatever reflects not this one Mind, is false and erroneous, even the belief that life, substance, and intelligence are both mental and material.

I Am. God; incorporeal and eternal Mind; divine Principle; the only Ego.

In. A term obsolete in Science if used with reference to Spirit, or Deity.

Intelligence. Substance; self-existent and eternal Mind; that which is never unconscious nor limited.

See chapter on Recapitulation, page 469.

Issachar (Jacob's son). A corporeal belief; the offspring of error; envy; hatred; selfishness; self-will; lust.

Jacob. A corporeal mortal embracing duplicity, repentance, sensualism. Inspiration; the revelation of Science, in which the so-called material senses yield to the spiritual sense of Life and Love.

Japhet (Noah's son). A type of spiritual peace, flowing from the understanding that God is the divine Principle of all existence, and that man is His idea, the child of His care.

Jerusalem. Mortal belief and knowledge obtained from the five corporeal senses; the pride of power and the power of pride; sensuality; envy; oppression; tyranny. Home, heaven.

Jesus. The highest human corporeal concept of the divine idea, rebuking and destroying error and bringing to light man's immortality.

Joseph. A corporeal mortal; a higher sense of Truth rebuking mortal belief, or error, and showing the immortality and supremacy of Truth; pure affection blessing its enemies.

Judah. A corporeal material belief progressing and disappearing; the spiritual understanding of God and man appearing.

Kingdom of Heaven. The reign of harmony in divine Science; the realm of unerring, eternal, and omnipotent Mind; the atmosphere of Spirit, where Soul is supreme.

Knowledge. Evidence obtained from the five corporeal senses; mortality; beliefs and opinions; human theories, doctrines, hypotheses; that which is not divine and is the origin of sin, sickness, and death; the opposite of spiritual Truth and understanding.

Lamb of God. The spiritual idea of Love; self-immolation; innocence and purity; sacrifice.

Levi (Jacob's son). A corporeal and sensual belief; mortal man; denial of the fulness of God's creation; ecclesiastical despotism.

Life. See chapter on Recapitulation, page 468.

Lord. In the Hebrew, this term is sometimes employed as a title, which has the inferior sense of master, or ruler. In the Greek, the word **kurios** almost always has this lower sense, unless specially coupled with the name God. Its higher signification is Supreme Ruler.

Lord God. Jehovah.

This double term is not used in the first chapter of Genesis, the record of spiritual creation. It is introduced in the second and following chapters, when the spiritual sense of God and of infinity is disappearing from the recorder's thought, - when the true scientific statements of the Scriptures become clouded through a physical sense of God as finite and corporeal. From this follow idolatry and mythology, - belief in many gods, or material intelligences, as the opposite of the one Spirit, or intelligence, named Elohim, or God.

Man. The compound idea of infinite Spirit; the spiritual image and likeness of God; the full representation of Mind.

Matter. Mythology; mortality; another name for mortal mind; illusion; intelligence, substance, and life in non-intelligence and mortality; life resulting in death, and death in life; sensation in the sensationless; mind originating in matter; the opposite of Truth; the opposite of Spirit; the opposite of God; that of which immortal Mind takes no cognizance; that which mortal mind sees, feels, hears, tastes, and smells only in belief.

Mind. The only I, or Us; the only Spirit, Soul, divine Principle, substance, Life, Truth, Love; the one God; not that which is **in** man, but the divine Principle, or God, of whom man is the full and perfect expression; Deity, which outlines but is not outlined.

Miracle. That which is divinely natural, but must be learned humanly; a phenomenon of

Science.

Morning. Light; symbol of Truth; revelation and progress.

Mortal Mind. Nothing claiming to be something, for Mind is immortal; mythology; error creating other errors; a suppositional material sense, **alias** the belief that sensation is in matter, which is sensationless; a belief that life, substance, and intelligence are in and of matter; the opposite of Spirit, and therefore the opposite of God, or good; the belief that life has a beginning and therefore an end; the belief that man is the offspring of mortals; the belief that there can be more than one creator; idolatry; the subjective states of error; material senses; that which neither exists in Science nor can be recognized by the spiritual sense; sin; sickness; death.

Moses. A corporeal mortal; moral courage; a type of moral law and the demonstration thereof; the proof that, without the gospel, - the union of justice and affection, - there is something spiritually lacking, since justice demands penalties under the law.

Mother. God; divine and eternal Principle; Life, Truth, and Love.

New Jerusalem. Divine Science; the spiritual facts and harmony of the universe; the kingdom of heaven, or reign of harmony.

Night. Darkness; doubt; fear.

Noah. A corporeal mortal; knowledge of the nothingness of material things and of the immortality of all that is spiritual.

Oil. Consecration; charity; gentleness; prayer; heavenly inspiration.

Pharisee. Corporeal and sensuous belief; self-righteousness; vanity; hypocrisy.

Pison (river). The love of the good and beautiful, and their immortality.

Principle. See chapter on Recapitulation, page 465.

Prophet. A spiritual seer; disappearance of material sense before the conscious facts of spiritual Truth.

Purse. Laying up treasures in matter; error.

Red Dragon. Error; fear; inflammation; sensuality; subtlety; animal magnetism; envy; revenge.

Resurrection. Spiritualization of thought; a new and higher idea of immortality, or spiritual existence; material belief yielding to spiritual understanding.

Reuben (Jacob's son). Corporeality; sensuality; delusion; mortality; error.

River. Channel of thought.

When smooth and unobstructed, it typifies the course of Truth; but muddy, foaming, and dashing, it is a type of error.

Rock. Spiritual foundation; Truth. Coldness and stubbornness.

Salvation. Life, Truth, and Love understood and demonstrated as supreme over all; sin, sickness, and death destroyed.

Seal. The signet of error revealed by Truth.

Serpent (**ophis**, in Greek; **nacash**, in Hebrew). Subtlety; a lie; the opposite of Truth, named error; the first statement of mythology and idolatry; the belief in more than one God; animal magnetism; the first lie of limitation; finity; the first claim that there is an opposite of Spirit, or good, termed matter, or evil; the first delusion that error exists as fact; the first claim that sin, sickness, and death are the realities of life. The first audible claim that God was not omnipotent and that there was another power, named **evil**, which was as real and eternal as God, good.

Sheep. Innocence; inoffensiveness; those who follow their leader.

Shem (Noah's son). A corporeal mortal; kindly affection; love rebuking error; reproof of sensualism.

Son. The Son of God, the Messiah or Christ. The son of man, the offspring of the flesh. "Son of a year."

Souls. See chapter on Recapitulation, page 466.

Spirit. Divine substance; Mind; divine Principle; all that is good; God; that only which is perfect, everlasting, omnipresent, omnipotent, infinite.

Spirits. Mortal beliefs; corporeality; evil minds; supposed intelligences, or gods; the opposites of God; errors; hallucinations. (See page 466.)

Substance. See chapter on Recapitulation, page 468.

Sun. The symbol of Soul governing man, - of Truth, Life, and Love.

Sword. The idea of Truth; justice. Revenge; anger.

Tares. Mortality; error; sin; sickness; disease; death.

Temple. Body; the idea of Life, substance, and intelligence; the superstructure of Truth; the shrine of Love; a material superstructure, where mortals congregate for worship.

Thummim. Perfection; the eternal demand of divine Science.

The Urim and Thummim, which were to be on Aaron's breast when he went before Jehovah, were holiness and purification of thought and deed, which alone can fit us for the office of spiritual teaching.

Time. Mortal measurements; limits, in which are summed up all human acts, thoughts, beliefs, opinions, knowledge; matter; error; that which begins before, and continues after, what is termed death, until the mortal disappears and spiritual perfection appears.

Tithe. Contribution; tenth part; homage; gratitude. A sacrifice to the gods.

Uncleanliness. Impure thoughts; error; sin; dirt.

Ungodliness. Opposition to the divine Principle and its spiritual idea.

Unknown. That which spiritual sense alone comprehends, and which is unknown to the material senses.

Paganism and agnosticism may define Deity as "the great unknowable;" but Christian Science brings God much nearer to man, and makes Him better known as the All-in-all, forever near.

Paul saw in Athens an altar dedicated "to the unknown God." Referring to it, he said to the Athenians: "Whom therefore ye ignorantly worship, Him declare I unto you." (Acts xvii. 23.)

Urim. Light.

The rabbins believed that the stones in the breast-plate of the high-priest had supernatural illumination, but Christian Science reveals Spirit, not matter, as the illuminator of all. The illuminations of Science give us a sense of the nothingness of error, and they show the spiritual inspiration of Love and Truth to be the only fit preparation for admission to the presence and power of the Most High.

Valley. Depression; meekness; darkness.

"Though I walk through the valley of the shadow of death, I will fear no evil." (Psalm xxiii. 4.)

Though the way is dark in mortal sense, divine Life and Love illumine it, destroy the unrest of mortal thought, the fear of death, and the supposed reality of error. Christian Science, contradicting sense, maketh the valley to bud and blossom as the rose.

Veil. A cover; concealment; hiding; hypocrisy.

The Jewish women wore veils over their faces in token of reverence and submission and in accordance with Pharisaical notions.

The Judaic religion consisted mostly of rites and ceremonies. The motives and affections of a man were of little value, if only he appeared unto men to fast. The great Nazarene, as meek as he was mighty, rebuked the hypocrisy, which offered long petitions for blessings upon material methods, but cloaked the crime, latent in thought, which was ready to spring into action and crucify God's anointed. The martyrdom of Jesus was the culminating sin of Pharisaism. It rent the veil of the temple. It revealed the false foundations and superstructures of superficial religion, tore from bigotry and superstition their coverings, and opened the sepulchre with divine Science, - immortality and Love.

Wilderness. Loneliness; doubt; darkness. Spontaneity of thought and idea; the vestibule in which a material sense of things disappears, and spiritual sense unfolds the great facts of existence.

Will. The motive-power of error; mortal belief; animal power. The might and wisdom of God.

"For this is the will of God." (I Thessalonians iv. 3.)

Will, as a quality of so-called mortal mind, is a wrongdoer; hence it should not be confounded with the term as applied to Mind or to one of God's qualities.

Wind. That which indicates the might of omnipotence and the movements of God's spiritual government, encompassing all things. Destruction; anger; mortal passions.

The Greek word for **wind** (pneuma) is used also for **spirit**, as in the passage in John's Gospel, the third chapter, where we read: "The wind [**pneuma**] bloweth where it listeth. . . . So is every one that is born of the Spirit **pneuma**]." Here the original word is the same in both cases, yet it has received different translations, as in other passages in this same chapter and elsewhere in the New Testament. This shows how our Master had constantly to employ words of material significance in order to unfold spiritual thoughts. In the record of Jesus' supposed death, we read: "He bowed his head, and gave up the ghost;" but this word **ghost** is **pneuma**. It might be translated **wind** or **air**, and the phrase is equivalent to our common statement, "He breathed his last." What Jesus gave up was indeed air, an etherealized form of matter, for never did he give up Spirit, or Soul.

Wine. Inspiration; understanding. Error; fornication; temptation; passion.

Year. A solar measurement of time; mortality; space for repentance.

"One day is with the Lord as a thousand years." (II Peter iii. 8.)

One moment of divine consciousness, or the spiritual understanding of Life and Love, is a foretaste of eternity. This exalted view, obtained and retained when the Science of being is understood, would bridge over with life discerned spiritually the interval of death, and man would be in the full consciousness of his immortality and eternal harmony, where sin, sickness, and death are unknown. Time is a mortal thought, the divisor of which is the solar year. Eternity is God's measurement of Soul-filled years.

You. As applied to corporeality, a mortal; finity.

Zeal. The reflected animation of Life, Truth, and Love. Blind enthusiasm; mortal will.

Zion. Spiritual foundation and superstructure; inspiration; spiritual strength. Emptiness; unfaithfulness; desolation.

The Manual of the Mother Church

of The First Church of Christ Scientist In Boston, Massachusesttes, by Mary Baker Eddy By-laws of the 88th Edition 1910

Tenets

of The Mother Church
The First Church of Christ, Scientist
To be signed by those uniting with The First Church of Christ, Scientist, in Boston, Mass.

1. As adherents of Truth, we take the inspired Word of the Bible as our sufficient guide to eternal Life.

2. We acknowledge and adore one supreme and infinite God. We acknowledge His Son, one Christ; the Holy Ghost or divine Comforter; and man in God's image and likeness.

3. We acknowledge God's forgiveness of sin in the destruction of sin and the spiritual understanding that casts out evil as unreal. But the belief in sin is punished so long as the belief lasts.

4. We acknowledge Jesus' atonement as the evidence of divine, efficacious Love, unfolding man's unity with God through Christ Jesus the Way-shower; (From *Science and Health with Key to the Scriptures* by Mary Baker Eddy).and we acknowledge that man is saved through Christ, through Truth, Life, and Love as demonstrated by the Galilean Prophet in healing the sick and overcoming sin and death.

5. We acknowledge that the crucifixion of Jesus and his resurrection served to uplift faith to understand eternal Life, even the allness of Soul, Spirit, and the nothingness of matter.

6. And we solemnly promise to watch, and pray for that Mind to be in us which was also in Christ Jesus; to do unto others as we would have them do unto us; and to be merciful, just, and pure.

MARY BAKER EDDY

Historical Sketch

In the spring of 1879, a little band of earnest seekers after Truth went into deliberations over forming a church without creeds, to be called the "Church of Christ, Scientist." They were members of evangelical churches, and students of Mrs. Mary Baker Eddy in Christian Science, and were known as "Christian Scientists."

At a meeting of the Christian Scientist Association, April 12, 1879, on motion of Mrs. Eddy, it was voted,-- To organize a church designed to commemorate the word and works of our Master, which should reinstate primitive Christianity and its lost element of healing.

Mrs. Eddy was appointed on the committee to draft the Tenets of The Mother Church--the chief corner stone whereof is, that Christian Science, as taught and demonstrated by our Master, casts out error, heals the sick, and restores the lost Israel: for "the stone which the builders rejected, the same is become the head of the corner."

The charter for the Church was obtained June, 1879, [*] and the same month the members, twentysix in number, extended a call to Mary Baker Eddy to become their pastor. She accepted the call, and was ordained A. D. 1881. Although walking through deep waters, the little Church went steadily on, increasing in numbers, and at every epoch saying,

"Hitherto hath the Lord helped us."

On the twenty-third day of September, 1892, at the request of Rev. Mary Baker Eddy, twelve of her students and Church members met and reorganized, under her jurisdiction, the Christian Science Church and named it, THE FIRST CHURCH OF CHRIST, SCIENTIST.

At this meeting twenty others of Mrs. Eddy's students and members of her former Church were elected members of this Church,--those with others that have since been elected were known as "First Members." The Church Tenets, Rules, and By-Laws, as prepared by Mrs. Eddy, were adopted. A By-Law adopted March 17, 1903, changed the title of "First Members" to "Executive Members." (On July 8, 1908, the By-Laws pertaining to "Executive Members" were repealed.) THE FIRST CHURCH OF CHRIST, SCIENTIST, IN BOSTON, MASS., is designed to be built on the Rock, Christ; even the understanding and demonstration of divine Truth, Life, and Love, healing and saving the world from sin and death; thus to reflect in some degree the Church Universal and Triumphant.

[*] Steps were taken to promote the Church of Christ, Scientist, in April, May and June; formal organization was accomplished and the charter obtained in August, 1879.

Church By-Laws

4ooo (1) CHURCH OFFICERS

Article I
NAMES, ELECTION, AND DUTIES

Names. SECTION 1. The Church officers shall consist of the Pastor Emeritus, a Board of Directors, a President, a Clerk, a Treasurer, and two Readers.

President. SECT. 2. The President shall be elected, subject to the approval of the Pastor Emeritus, by the Board of Directors [1] on Monday preceding the annual meeting of the Church. The President shall hold office for one year, and the same person is eligible for election but once in three years.
[1] See under "Deed of Trust" for incorporation of the "Christian Science Board of Directors."

Clerk and Treasurer. SECT. 3. The term of office for the Clerk and the Treasurer of this Church (also for the editors and the manager of The Christian Science Publishing Society, and the manager of the general Committee on Publication in Boston) is one year each, dating from the time of election to office. Incumbents who have served one year or more, may be reelected, or new officers elected, at the annual meeting held for this purpose, by a unanimous vote of the Christian Science Board of Directors and the consent of the Pastor Emeritus given in her own handwriting.

Readers. SECT. 4. Every third year Readers shall be elected in The Mother Church by the Board of Directors, which shall inform the Pastor Emeritus of the names of its candidates before they are elected; and if she objects, said candidates shall not be chosen. The Directors shall fix the salaries of the Readers.

Directors. SECT. 5. The Christian Science Board of Directors shall consist of five members. They shall fill a vacancy occurring on that Board after the candidate is approved by the Pastor Emeritus. A majority vote or the request of Mrs. Eddy shall dismiss a member. Members shall neither report the discussions of this Board, nor those with Mrs. Eddy.

Church Business. SECT. 6. The business of The Mother Church shall be transacted by its Christian Science Board of Directors. The manager of the general Committee on Publication in the United States shall order no special action to be taken by said Committee that is not named in the Manual of this Church without consulting with the full Board of Directors of The Mother Church and receiving the written consent of said Board.

Publishing Buildings. SECT. 7. It shall be the duty of the Christian Science Board of Directors to provide a suitable building for the publication of *The Christian Science Journal, Christian Science Sentinel, Der Herold der Christian Science,* and all other Christian Science literature published by The Christian Science Publishing Society. It shall also be the duty of the Christian Science Board of Directors to provide suitable rooms, conveniently and pleasantly located in the same building, for the publication and sale of the books of which Mary Baker Eddy is, or may be, the author, and of other literature connected therewith.

Trusteeships and Syndicates. SECT. 8. Boards of Trustees and Syndicates may be formed by The Mother Church, subject to the approval of the Pastor Emeritus.

Duties of Church Officers. SECT. 9. Law constitutes government, and disobedience to the laws of The Mother Church must ultimate in annulling its Tenets and By-Laws. Without a proper system of government and form of action, nations, individuals, and religion are unprotected; hence the necessity of this By-Law and the warning of Holy Writ: "That servant, which knew his lord's will, and prepared not himself, neither did according to his will, shall be beaten with many stripes."

It is the duty of the Christian Science Board of Directors to watch and make sure that the officers of this Church perform the functions of their several offices promptly and well. If an officer fails to fulfil all the obligations of his office, the Board of Directors shall immediately call a meeting and notify this officer either to resign his place or to perform his office faithfully; then failing to do either, said officer shall be dismissed from this Church, and his dismissal shall be written on the Church records.

It is the duty of any member of this Church, and especially of one who has been or who is the First Reader of a church, to inform the Board of Directors of the failure of the Committee on Publication or of any other officer in this Church to perform his official duties. A Director shall not make known the name of the complainant.

If the Christian Science Board of Directors fails to fulfil the requirements of this By-Law, and a member of this Church or the Pastor Emeritus shall complain thereof to the Clerk and the complaint be found valid, the Directors shall resign
their office or perform their functions faithfully. Failing to do thus, the Pastor Emeritus shall appoint five suitable members of this Church to fill the vacancy. The salary of the members of the Board of Directors shall be at present two thousand five hundred dollars each annually.

Article II
READERS OF THE MOTHER CHURCH

Election. SECTION 1. The Readers for The Mother Church shall be a man and a woman, one to read the BIBLE, and one to read SCIENCE AND HEALTH WITH KEY TO THE SCRIPTURES.

Eligibility. SECT. 2. The Directors shall select intelligible Readers who are exemplary Christians and good English scholars. They must be members of The Mother Church.

Removal. SECT. 3. If a Reader in The Mother Church be found at any time inadequate or unworthy, he or she shall be removed from office by a majority vote of the Board of Directors and the consent of the Pastor Emeritus, and the vacancy supplied.

First Reader's Residence. SECT. 4. Unless Mrs. Eddy requests otherwise, the First Reader of The Mother Church shall occupy, during his term of Readership, the house of the Pastor Emeritus, No. 385 Commonwealth Avenue, Boston. The Board of Directors shall pay from the Church funds the taxes and rent on this property; the Board shall attend to the insurance before it expires, suitably furnish the house, and keep the property in good repair, so long as Mrs. Eddy does not occupy the house herself and the occupants are satisfactory to her.

Article III
DUTIES OF READERS OF THE MOTHER CHURCH AND OF ITS BRANCH CHURCHES

Moral Obligations. SECTION 1. The Readers of The Mother Church and of all its branch churches must devote a suitable portion of their time to preparation for the reading of the Sunday lesson,--a lesson on which the prosperity of Christian Science largely depends. They must keep themselves unspotted from the world,-- uncontaminated with evil,--that the mental atmosphere they exhale shall promote health and holiness, even that spiritual *animus* so universally needed.

First Readers' Duties. SECT. 2. It shall be the duty of the First Readers to conduct the principal part of the Sunday services, and the Wednesday evening meetings.

Suitable Selections. SECT. 3. The First Readers shall read, as a part of the Wednesday evening services, selections from the SCRIPTURES, and from SCIENCE AND HEALTH WITH KEY TO THE SCRIPTURES.

Order of Reading. SECT. 4. The First Readers in the Christian Science churches shall read the correlative texts in SCIENCE AND HEALTH WITH KEY TO THE SCRIPTURES; and the Second Readers shall read the BIBLE texts. The readings from the SCRIPTURES shall precede the readings from SCIENCE AND HEALTH. The Readers shall not read from copies or manuscripts, but from the books.

Naming Book and Author. SECT. 5. The Readers of SCIENCE AND HEALTH WITH KEY TO THE SCRIPTURES, before commencing to read from this book, shall distinctly announce the full title of the book and give the author's name. Such announcement shall be made but once during the lesson.

Readers in Branch Churches. SECT. 6. These Readers shall be members of The Mother Church. They shall read understandingly and be well educated. They shall make no remarks explanatory of the LESSON-SERMON at any time, but they shall read all notices and remarks that may be printed in the CHRISTIAN SCIENCE QUARTERLY. This By-Law applies to Readers in all the branch churches.

Enforcement of By-Laws. SECT. 7. It shall be the duty of every member of The Mother Church, who is a First Reader in a Church of Christ, Scientist, to enforce the discipline and by-laws of the church in which he is Reader.

A Reader not a Leader. SECT. 8. The Church Reader shall not be a Leader, but he shall maintain the Tenets, Rules, and discipline of the Church. A Reader shall not be a President of a church.

3ooo (2) CHURCH MEMBERSHIP

Article IV
QUALIFICATIONS FOR MEMBERSHIP

Believe in Christian Science. SECTION 1. To become a member of The Mother Church, The First Church of Christ, Scientist, in Boston, Mass., the applicant must be a believer in the doctrines of Christian Science, according to the platform and teaching contained in the Christian Science textbook, SCIENCE AND HEALTH WITH KEY TO THE SCRIPTURES, by Rev. Mary Baker Eddy. The BIBLE, together with SCIENCE AND HEALTH and other works by Mrs. Eddy, shall be his only textbooks for self-instruction in Christian Science, and for teaching and practising metaphysical healing.

Free from Other Denominations. SECT. 2. This Church will receive a member of another Church of Christ, Scientist, but not a church member from a different denomination until that membership is dissolved.

Children when Twelve Years Old. SECT. 3. Children who have arrived at the age of twelve years, who are approved, and whose applications are countersigned by one of Mrs. Eddy's loyal students, by a Director, or by a student of the Board of Education, may be admitted to membership with The Mother Church.

Article V
APPLICATIONS FOR MEMBERSHIP

Students of the College. SECTION 1. Applications for membership with The Mother Church from students of the Massachusetts Metaphysical College who studied with Rev. Mary Baker Eddy, shall be signed by the Christian Science Board of Directors as evidence of the loyalty of the applicants.

Other Students. SECT. 2. Applicants for membership who have not studied Christian Science with Rev. Mary Baker Eddy, can unite with this Church only by approval from students of Mrs. Eddy, loyal to the teachings of the textbook, SCIENCE AND HEALTH WITH KEY TO THE SCRIPTURES, or from members of The Mother Church, as provided in Article VI, Sect. 2, of these By-Laws.

Students' Pupils. SECT. 3. Applications for membership with The Mother Church, coming from pupils of loyal students who have taken the Primary or Normal Course at the Massachusetts Metaphysical College or in the Board of Education, or from pupils of those who have passed an examination by the Board of Education, shall have the approval and signature of their teachers, except in such cases as are provided for in Sect. 4 of this Article.

Exceptional Cases. SECT. 4. Loyal Christian Scientists whose teachers are deceased, absent, or disloyal,--or those whose teachers, for insufficient cause, refuse to endorse their applications for membership with The Mother Church, --can apply to the Clerk of this Church, and present to him a recommendation signed by three members thereof in good standing, after which, the unanimous vote of the Board of Directors may admit said applicant to membership.

Addressed to Clerk. SECT. 5. All applications for membership must be addressed to the Clerk of the Church.

Endorsing Applications. SECT. 6. A member of The Mother Church shall not endorse nor countersign an application for membership therewith until after the blank has been properly filled out by an applicant. A member who violates this By-Law shall be disciplined.

Notice of Rejection. SECT. 7. If an application for membership with The First Church of Christ, Scientist, in Boston, Mass., is rejected, the Clerk of the Church shall send to the applicant a notice of such rejection; but neither the Clerk nor the Church shall be obliged to report the cause for rejection.

Article VI
RECOMMENDATION AND ELECTION

Pupils of Normal Students. SECTION 1. One Normal student cannot recommend the pupil of another Normal student, so long as both are loyal to their Leader and to the Christian Science textbook, except as provided for in Article V, Sect. 4.

Members of The Mother Church. SECT. 2. Only members of The Mother Church are qualified to approve for membership individuals who are known to them to be Christians, and faithful, loyal students of the textbook, SCIENCE AND HEALTH WITH KEY TO THE SCRIPTURES. If the approver is not a loyal student of Mrs. Eddy, a Director of this Church, or a student of the Board of Education who holds a degree, the application must be countersigned by one of these.

Election. SECT. 3. Applicants for membership in this Church, whose applications are correctly prepared, may be elected by majority vote of the Christian Science Board of Directors at the semiannual meetings held for this purpose.

Article VII
PROBATIONARY MEMBERSHIP

Members who once Withdrew. SECTION 1. Individuals who have heretofore been members of this Church, or were members of the Church of Christ, Scientist, organized in 1879 by Mary Baker Eddy, but who have voluntarily withdrawn, may be received into this Church on one year's probation, provided they are willing and anxious to live according to its requirements and make application for membership according to its By-Laws. If, at the expiration of said one year, they are found worthy, they shall be received into full membership, but if not found worthy their applications shall be void.

Members once Dismissed. SECT. 2. A full member or a probationary member, who has been excommunicated once, and who afterward, when sufficient time has elapsed thoroughly to test his sincerity, gives due evidence of having genuinely repented and of being radically reformed, shall be eligible to probationary membership upon a unanimous vote of the Christian Science Board of Directors.

Ineligible for Probation. SECT. 3. If a member has been twice notified of his excommunication, he shall not again be received into this Church.

2ooo (3) DISCIPLINE

Article VIII
GUIDANCE OF MEMBERS

A Rule for Motives and Acts. SECTION 1. Neither animosity nor mere personal attachment should impel the motives or acts of the members of The Mother Church. In Science, divine Love alone governs man; and a Christian Scientist reflects the sweet amenities of Love, in rebuking sin, in true brotherliness, charitableness, and forgiveness. The members of this Church should daily watch and pray to be delivered from all evil, from prophesying, judging, condemning, counseling, influencing or being influenced erroneously.

To be Read in Church. SECT. 2. The above Church Rule shall be read in The Mother Church and in the branch churches by the First Reader on the first Sunday of each month. On Communion day the Church Tenets are to be read.

Christ Jesus the Ensample. SECT. 3. He who dated the Christian era is the Ensample in Christian Science. Careless comparison or irreverent reference to Christ Jesus is abnormal in a Christian Scientist, and is prohibited. When it is necessary to show the great gulf between Christian Science and theosophy, hypnotism, or spiritualism, do it, but without hard words. The wise man saith, "A soft answer turneth away wrath." However despitefully used and misrepresented by the churches or the press, in return employ no violent invective, and do good unto your enemies when the opportunity occurs. A departure from this rule disqualifies a member for office in the Church or on the Board of Lectureship, and renders this member liable to discipline and, possibly, dismissal from The Mother Church.

Daily Prayer. SECT. 4. It shall be the duty of every member of this Church to pray each day: "Thy kingdom come;" let the reign of divine Truth, Life, and Love be established in me, and rule out of me all sin; and may Thy Word enrich the affections of all mankind, and govern them!

Prayer in Church. SECT. 5. The prayers in Christian Science churches shall be offered for the congregations collectively and exclusively.

Alertness to Duty. SECT. 6. It shall be the duty of every member of this Church to defend himself daily against aggressive mental suggestion, and not be made to forget nor to neglect his duty to God, to his Leader, and to mankind. By his works he shall be judged,--and justified or condemned.

One Christ. SECT. 7. In accordance with the Christian Science textbooks,--the BIBLE, and SCIENCE AND HEALTH WITH KEY TO THE SCRIPTURES,--and in accord with all of Mrs. Eddy's teachings, members of this Church shall neither entertain a belief nor signify a belief in more than one Christ, even that Christ whereof the Scripture beareth testimony.

No Malpractice. SECT. 8. Members will not intentionally or knowingly mentally malpractise, inasmuch as Christian Science can only be practised according to the Golden Rule: "All things whatsoever ye would that men should do to you, do ye even so to them." (Matt. 7:12.)

A member of The Mother Church who mentally malpractises upon or treats our Leader or her staff without her or their consent shall be disciplined, and a second offense as aforesaid shall cause the name of said member to be dropped forever from The Mother Church.

Formulas Forbidden. SECT. 9. No member shall use written formulas, nor permit his patients or pupils to use them, as auxiliaries to teaching Christian Science or for healing the sick. Whatever is requisite for either is contained in the books of the Discoverer and Founder of Christian Science. Sometimes she may strengthen the faith by a written text as no one else can.

No Adulterating Christian Science. SECT. 10. A member of this Church shall not publish profuse quotations from Mary Baker Eddy's copyrighted works without her permission, and shall not plagiarize her writings. This By-Law not only calls more serious attention to the commandment of the Decalogue, but tends to prevent Christian Science from being *adulterated*.

No Incorrect Literature. SECT. 11. A member of this Church shall neither buy, sell, nor circulate Christian Science literature which is not correct in its statement of the divine Principle and rules and the demonstration of Christian Science. Also the spirit in which the writer has written his literature shall be definitely considered. His writings must show strict adherence to the Golden Rule, or his literature shall not be adjudged Christian Science. A departure from the spirit or letter of this By-Law involves schisms in our Church and the possible loss, for a time, of Christian Science.

Obnoxious Books. SECT. 12. A member of this Church shall not patronize a publishing house or bookstore that has for sale obnoxious books.

Per Capita Tax. SECT. 13. Every member of The Mother Church shall pay annually a per capita tax of not less than one dollar, which shall be forwarded each year to the Church Treasurer.

Church Periodicals. SECT. 14. It shall be the privilege and duty of every member, who can afford it, to subscribe for the periodicals which are the organs of this Church; and it shall be the duty of the Directors to see that these periodicals are ably edited and kept abreast of the times.

Church Organizations Ample. SECT. 15. Members of this Church shall not unite with organizations which impede their progress in Christian Science. God requires our whole heart, and He supplies within the wide channels of The Mother Church dutiful and sufficient occupation for all its members.

Joining Another Society. SECT. 16. It shall be the duty of the members of The Mother Church and of its branches to promote peace on earth and good will toward men; but members of The Mother Church shall not hereafter become members of other societies except those specified in The Mother Church Manual, and they shall strive to promote the welfare of all mankind by demonstrating the rules of divine Love.

Forbidden Membership. SECT. 17. A member of The First Church of Christ, Scientist, in Boston, Mass., shall not be a member of any church whose Readers are not Christian Scientists and members of The Mother Church.

Officious Members. SECT. 18. A member of The Mother Church is not entitled to hold office or read in branch churches of this denomination except by invitation.

Legal Titles. SECT. 19. Students of Christian Science must drop the titles of Reverend and Doctor, except those who have received these titles under the *laws* of the *State*.

Illegal Adoption. SECT. 20. No person shall be a member of this Church who claims a spiritually adopted child or a spiritually adopted husband or wife. There must be legal adoption and legal marriage, which can be verified according to the laws of our land.

Use of Initials "C.S." SECT. 21. A member of The Mother Church shall not place the initials "C.S." after his name on circulars, cards, or leaflets, which advertise his business or profession, except as a Christian Science practitioner.

Practitioners and Patients. SECT. 22. Members of this Church shall hold in sacred confidence all private communications made to them by their patients; also such information as may come to them by reason of their relation of practitioner to patient. A failure to do this shall subject the offender to Church discipline.

A member of The Mother Church shall not, under pardonable circumstances, sue his patient for recovery of payment for said member's practice, on penalty of discipline and liability to have his name removed from membership. Also he shall reasonably reduce his price in chronic cases of recovery, and in cases where he has not effected a cure. A Christian Scientist is a humanitarian; he is benevolent, forgiving, long-suffering, and seeks to overcome evil with good.

Duty to Patients. SECT. 23. If a member of this Church has a patient whom he does not heal, and whose case he cannot fully diagnose, he may consult with an M.D. on the anatomy involved. And it shall be the privilege of a Christian Scientist to confer with an M.D. on Ontology, or the Science of being.

Testimonials. SECT. 24. "Glorify God in your body, and in your spirit, which are God's" (St. Paul). Testimony in regard to the healing of the sick is highly important. More than a mere rehearsal of blessings, it scales the pinnacle of praise and illustrates the demonstration of Christ, "who healeth all thy diseases" (Psalm 103:3). This testimony, however, shall not include a description of symptoms or of suffering, though the generic name of the disease may be indicated. This By-Law applies to testimonials which appear in the periodicals and to those which are given at the Wednesday evening meeting.

Charity to All. SECT. 25. While members of this Church do not believe in the doctrines of theosophy, hypnotism, or spiritualism, they cherish no enmity toward those who do believe in such doctrines, and will not harm them. But whenever God calls a member to bear testimony to Truth and to defend the Cause of Christ, he shall do it with love and without fear.

Uncharitable Publications. SECT. 26. A member of this Church shall not publish, nor cause to be published, an article that is uncharitable or impertinent towards religion, medicine, the courts, or the laws of our land.

The Golden Rule. SECT. 27. A member of The Mother Church shall not haunt Mrs. Eddy's drive when she goes out, continually stroll by her house, or make a summer resort near her for such a purpose. Numbering the People. SECT. 28. Christian Scientists shall not report for publication the number of the members of The Mother Church, nor that of the branch churches. According to the Scripture they shall turn away from personality and numbering the people.

Our Church Edifices. SECT. 29. The periodicals of our denomination do not publish descriptions of our church edifices, but they may quote from other periodicals or give incidental narratives.

No Monopoly. SECT. 30. A Scientist shall not endeavor to monopolize the healing work in any church or locality, to the exclusion of others, but all who understand the teachings of Christian Science are privileged to enter into this holy work, and "by their fruits ye shall know them."

Christian Science Nurse. SECT. 31. A member of The Mother Church who represents himself or herself as a Christian Science nurse shall be one who has a demonstrable knowledge of Christian Science practice, who thoroughly understands the practical wisdom necessary in a sick room, and who can take proper care of the sick. The cards of such persons may be inserted in *The Christian Science Journal* under rules established by the publishers.

Article IX
MARRIAGE AND DECEASE

A Legal Ceremony. SECTION 1. If a Christian Scientist is to be married, the ceremony shall be performed by a clergyman who is legally authorized.

Sudden Decease. SECT. 2. If a member of The Mother Church shall decease suddenly, without previous injury or illness, and the cause thereof be unknown, an autopsy shall be made by qualified experts. When it is possible the body of a female shall be prepared for burial by *one of her own sex*.

Article X
DEBATING IN PUBLIC

No Unauthorized Debating. SECTION 1. A member of this Church shall not debate on Christian Science in public debating assemblies, without the consent of the Board of Directors.

Article XI
COMPLAINTS

Departure from Tenets. SECTION 1. If a member of this Church shall depart from the Tenets and be found having the name without the life of a Christian Scientist, and another member in good standing shall from Christian motives make this evident, a meeting of the Board of Directors shall be called, and the offender's case shall be tried and said member exonerated, put on probation, or excommunicated.

Violation of By-Laws. SECT. 2. A member who is found violating any of the By-Laws or Rules herein set forth, shall be admonished in consonance with the Scriptural demand in Matthew 18:15-17; and if he neglect to accept such admonition, he shall be placed on probation, or if he repeat the offense, his name shall be dropped from the roll of Church membership.

Violation of Christian Fellowship. SECT. 3. Any member who shall unjustly aggrieve or vilify the Pastor Emeritus or another member, or who does not live in Christian fellowship with members who are in good and regular standing with this Church, shall either withdraw from the Church or be excommunicated.

Preliminary Requirement. SECT. 4. No Church discipline shall ensue until the requirements according to the Scriptures, in Matthew 18:15-17, have been strictly obeyed, unless a By-Law governing the case provides for immediate action.

Authority. SECT. 5. The Christian Science Board of Directors has power to discipline, place on probation, remove from membership, or to excommunicate members of The Mother Church. Only the members of this Board shall be present at meetings for the examination of complaints against Church members; and they alone shall vote on cases involving The Mother Church discipline.

Members in Mother Church Only. SECT. 6. A complaint against a member of The Mother Church, *if said member belongs to no branch church* and if this complaint is not for *mental malpractice*, shall be laid before this Board, and within ten days thereafter, the Clerk of the Church shall address a letter of inquiry to the member complained of as to the validity of the charge. If a member is found guilty of that whereof he is accused and his previous character has been good, his confession of his error and evidence of his compliance with our Church Rules shall be deemed sufficient by the Board for forgiveness for once, and the Clerk of the Church shall immediately so inform him. But a second offense shall dismiss a member from the Church.

Working Against the Cause. SECT. 7. If a member of this Church shall, mentally or otherwise, persist in working against the interests of another member, or the interests of our Pastor Emeritus and the accomplishment of what she understands is advantageous to this Church and to the Cause of Christian Science, or shall influence others thus to act, upon her complaint or the complaint of a member for her or for himself, it shall be the duty of the Board of Directors immediately to call a meeting, and drop forever the name of the member guilty of this offense from the roll of Church membership.

No Unchristian Conduct. SECT. 8. If a member of this Church were to treat the author of our textbook disrespectfully and cruelly, upon her complaint that member should be excommunicated. If a member, without her having requested the information, shall trouble her on subjects unnecessarily and without her consent, it shall be considered an offense.

Not to Learn Hypnotism. SECT. 9. Members of this Church shall not learn hypnotism on penalty of being excommunicated from this Church. No member shall enter a complaint of mental malpractice for a sinister purpose. If the author of SCIENCE AND HEALTH shall bear witness to the offense of mental malpractice, it shall be considered a sufficient evidence thereof.

Publications Unjust. SECT. 10. If a member of The Mother Church publishes, or causes to be published, an article that is false or unjust, hence injurious, to Christian Science or to its Leader, and if, upon complaint by another member, the Board of Directors finds that the offense has been committed, the offender shall be suspended for not less than three years from his or her office in this Church and from Church membership.

The Mother Church of Christ, Scientist, Tenets. SECT. 11. If a member of The Mother Church of Christ, Scientist, or a member of a branch of this Church break the rules of its Tenets as to unjust and unmerciful conduct--on complaint of Mrs. Eddy our Pastor Emeritus --and this complaint being found valid, his or her name shall be erased from The Mother Church and the branch church's list of membership and the offender shall not be received into The Mother Church or a branch church for twelve years.

Special Offense. SECT. 12. If a member of this Church, either by word or work, represents falsely to or of the Leader and Pastor Emeritus, said member shall immediately be disciplined, and a second similar offense shall remove his or her name from membership in The Mother Church.

Members of Branch Churches. SECT. 13. A member of both The Mother Church and a branch Church of Christ, Scientist, or a Reader, shall not report nor send notices to The Mother Church, or to the Pastor Emeritus, of errors of the members of their local church; but they shall strive to overcome these errors. Each church shall separately and independently discipline its own members,--if this sad necessity occurs.

Article XII
TEACHERS

Probation. SECTION 1. For sufficient reasons it may be decided that a teacher has so strayed as not to be fit for the work of a Reader in church or a teacher of Christian Science. Although repentant and forgiven by the Church and retaining his membership, this weak member shall not be counted loyal till after three years of exemplary character. Then the Board of Directors may decide if his loyalty has been proved by uniform maintenance of the life of a consistent, consecrated Christian Scientist.

Misteaching. SECT. 2. If a member of this Church is found trying to practise or to teach Christian Science contrary to the statement thereof in its textbook, SCIENCE AND HEALTH WITH KEY TO THE SCRIPTURES, it shall be the duty of the Board of Directors to admonish that member according to Article XI, Sect. 4. Then, if said member persists in this offense, his or her name shall be dropped from the roll of this Church.

1ooo (4) MEETINGS

Article XIII
REGULAR AND SPECIAL MEETINGS

Annual Meetings. SECTION 1. The regular meetings of The Mother Church shall be held annually, on Monday following the first Sunday in June. No other than its officers are required to be present. These assemblies shall be for listening to the reports of Treasurer, Clerk, and Committees, and general reports from the Field.

Meetings of Board of Directors. SECT. 2. The annual meeting of the Christian Science Board of Directors, for electing officers and other business, shall be held on Monday preceding the annual meeting of the Church. Regular meetings for electing candidates to membership with The Mother Church, and for the transaction of such other business as may properly come before these meetings, shall be held on the Friday preceding the first Sunday in June, and on the first Friday in November of each year. Special meetings may be held at any time upon the call of the Clerk.

Called only by the Clerk. SECT. 3. Before calling a meeting of the members of this Church (excepting its regular sessions) it shall be the duty of the Clerk to inform the Board of Directors and the Pastor Emeritus of his intention, and to state definitely the purpose for which the members are to convene. The Clerk must have the consent of this Board and the Pastor Emeritus, before he can call said meeting.

o4oo (5) CHURCH SERVICES

Article XIV
THE CHRISTIAN SCIENCE PASTOR

Ordination. SECTION 1. I, Mary Baker Eddy, ordain the BIBLE, and SCIENCE AND HEALTH WITH KEY TO THE SCRIPTURES, Pastor over The Mother Church,--The First Church of Christ, Scientist, in Boston, Mass.,--and they will continue to preach for this Church and the world.

The Lesson-Sermon. SECT. 2. The subject of the Lesson-Sermon in the morning service of The Mother Church, and of the branch Churches of Christ, Scientist, shall be repeated at the other services on Sunday. The correlative Biblical texts in the Lesson Sermon shall extend from Genesis to Revelation.

Article XV
READING IN PUBLIC

Announcing Author's Name. SECTION 1. To pour into the ears of listeners the sacred revelations of Christian Science indiscriminately, or without characterizing their origin and thus distinguishing them from the writings of authors who think at random on this subject, is to lose some weight in the scale of right thinking. Therefore it is the duty of every member of this Church, when publicly reading or quoting from the books or poems of our Pastor Emeritus, first to announce the name of the author. Members shall also instruct their pupils to adopt the aforenamed method for the benefit of our Cause.

Article XVI
WELCOMING STRANGERS

The Leader's Welcome. SECTION 1. Mrs. Eddy welcomes to her seats in the church, persons of all sects and denominations who come to listen to the Sunday sermon and are not otherwise provided with seats.

The Local Members' Welcome. SECT. 2. It shall be the duty and privilege of the local members of The Mother Church to give their seats, if necessary, to strangers who may come to attend the morning services.

Article XVII
SERVICES UNINTERRUPTED

Continued Throughout the Year. SECTION 1. The services of The Mother Church shall be continued twelve months each year. One meet-ing on Sunday during the months of July and August is sufficient. A Christian Scientist is not fatigued by prayer, by reading the Scriptures or the Christian Science textbook. Amusement or idleness is weariness. Truth and Love rest the weary and heavy laden.

Easter Observances. SECT. 2. In the United States there shall be no special observances, festivities, nor gifts at the Easter season by members of The Mother Church. Gratitude and love should abide in every heart each day of all the years. Those sacred words of our beloved Master, "Let the dead bury their dead," and "Follow thou me," appeal to daily Christian endeavors for the living whereby to exemplify our risen Lord.

Laying a Corner Stone. SECT. 3. No large gathering of people nor display shall be allowed when laying the Corner Stone of a Church of Christ, Scientist. Let the ceremony be devout. No special trowel should be used. (See SCIENCE AND HEALTH, page 140.)

Overflow Meetings. SECT. 4. A Church of Christ, Scientist, shall not hold two or more Sunday services at the same hour.

Article XVIII
COMMUNION

No more Communion. SECTION 1. The Mother Church of Christ, Scientist, shall observe no more Communion seasons.

Communion of Branch Churches. SECT. 2. The Communion shall be observed in the branch churches on the second Sunday in January and July of each year, and at this service the Tenets of The Mother Church are to be read.

Article XIX
MUSIC IN THE CHURCH

Soloist and Organist. SECTION 1. The music in The Mother Church shall not be operatic, but of an appropriate religious character and of a recognized standard of musical excellence; it shall be played in a dignified and suitable manner. Music from the organ alone should continue about eight or nine minutes for the voluntary and six or seven minutes for the postlude, the offertory conforming to the time required to take the collection. The solo singer shall not neglect to sing any special hymn selected by the Board of Directors.

Article XX
SUNDAY SCHOOL

The Sunday School. SECTION 1. Pupils may be received in the Sunday School classes of any Church of Christ, Scientist, up to the age of twenty years, and by transfer from another Church of Christ, Scientist, up to that age, but no pupil shall remain in the Sunday School of any Church of Christ, Scientist, after reaching the age of twenty. None except the officers, teachers, and pupils should attend the Sunday School exercises.

Teaching the Children. SECT. 2. The Sabbath School children shall be taught the Scriptures, and they shall be instructed according to their understanding or ability to grasp the simpler meanings of the divine Principle that they are taught.

Subject for Lessons. SECT. 3. The first lessons of the children should be the Ten Commandments (Exodus 20:3-17), the Lord's Prayer (Matt. 6:9-13), and its Spiritual Interpretation by Mary Baker Eddy, Sermon on the Mount (Matt. 5:3-12). The next lessons consist of such questions and answers as are adapted to a juvenile class, and may be found in the Christian Science Quarterly Lessons, read in Church services. The instruction given by the children's teachers must not deviate from the absolute Christian Science contained in their textbook.

o3oo (6) READING ROOMS

Article XXI

Establishment. SECTION 1. Each church of the Christian Science denomination shall have a Reading Room, though two or more churches may unite in having Reading Rooms, provided these rooms are well located.

Librarian. SECT. 2. The individuals who take charge of the Reading Rooms of The Mother Church shall be elected by the Christian Science Board of Directors, subject to the approval of Mary Baker Eddy. He or she shall have no bad habits, shall have had experience in the Field, shall be well educated, and a devout Christian Scientist. (See also Article XXV, Sect. 7.)

Literature in Reading Rooms. SECT. 3. The literature sold or exhibited in the Reading Rooms of Christian Science Churches shall consist only of
Science and Health with Key to the Scriptures, by Mary Baker Eddy, and other writings by this author; also the literature published or sold by The Christian Science Publishing Society.

o2oo (7) RELATION AND DUTIES OF MEMBERS TO PASTOR EMERITUS

Article XXII

The Title of Mother Changed. SECTION 1. In the year eighteen hundred and ninety-five, loyal Christian Scientists had given to the author of their textbook, the Founder of Christian Science, the individual, endearing term of Mother. At first Mrs. Eddy objected to being called thus, but afterward consented on the ground that this appellative in the Church meant nothing more than a tender term such as sister or brother. In the year nineteen hundred and three and after, owing to the public misunderstanding of this name, it is the duty of Christian Scientists to drop the word *mother* and to substitute Leader, already used in our periodicals.

A Member not a Leader. SECT. 2. A member of The First Church of Christ, Scientist, in Boston, Mass., shall not be called Leader by members of this Church, when this term is used in connection with Christian Science.

Obedience Required. SECT. 3. It shall be the duty of the officers of this Church, of the editors of the *Christian Science Journal, Sentinel*, and *Der Herold*, of the members of the Committees on Publication, of the Trustees of The Christian Science Publishing Society, and of the Board of Education promptly to comply with any written order, signed by Mary Baker Eddy, which applies to their official functions. Disobedience to this By-Law shall be sufficient cause for the removal of the offending member from office.

The vacancy shall be supplied by a majority vote of the Christian Science Board of Directors, and the candidate shall be subject to the approval of Mary Baker Eddy.

Understanding Communications. SECT. 4. If the Clerk of this Church shall receive a communication from the Pastor Emeritus which he does not fully understand, he shall inform her of this fact before presenting it to the Church and obtain a clear understanding of the matter, --then act in accordance therewith.

Interpreting Communications. SECT. 5. If at a meeting of this Church a doubt or disagreement shall arise among the members as to the signification of the communications of the Pastor Emeritus to them, before action is taken it shall be the duty of the Clerk to report to her the vexed question and to await her explanation thereof.

Reading and Attesting Letters. SECT. 6. When a letter or a message from the Pastor Emeritus is brought before a meeting of this Church, or she is referred to as authority for business, it shall be the duty of the Church to inquire if all of the letter has been read, and to require all of it to be read; also to have any authority supposed to come from her satisfactorily attested.

Unauthorized Reports. SECT. 7. Members of this Church shall not report on authority an order from Mrs. Eddy that she has not sent, either to the Boards or to the executive bodies of this Church. The Pastor Emeritus is not to be consulted on cases of discipline, on the cases of candidates for admission to this Church, or on the cases of those on trial for dismissal from the Church.

Private Communications. SECT. 8. A strictly private communication from the Pastor Emeritus to a member of her Church shall not be made public without her written consent.

Unauthorized Legal Action. SECT. 9. A member of this Church shall not employ an attorney, nor take legal action on a case not provided for in its By-Laws-if said case relates to the person or to the property of Mary Baker Eddy--without having personally conferred with her on said subject.

Duty to God. SECT. 10. Members of this Church who turn their attention from the divine Principle of being to personality, sending gifts, congratulatory despatches or letters to the Pastor Emeritus on Thanksgiving, Christmas, New Year, or Easter, break a rule of this Church and are amenable therefor.

Opportunity for Serving the Leader. SECT. 11. At the written request of the Pastor Emeritus, Mrs. Eddy, the Board of Directors shall immediately notify a person who has been a member of this Church at least three years to go in ten days to her, and it shall be the duty of the member thus notified to remain with Mrs. Eddy three years consecutively. A member who leaves her in less time without the Directors' consent or who declines to obey this call to duty, upon Mrs. Eddy's complaint thereof shall be excommunicated from The Mother Church. Members thus serving the Leader shall be paid semi-annually at the rate of one thousand dollars yearly in addition to rent and board. Those members whom she teaches the course in Divinity, and who remain with her three consecutive years, receive the degree of the Massachusetts Metaphysical College.

Location. SECT. 12. Rev. Mary Baker Eddy calls to her home or allows to visit or to locate therein only those individuals whom she engages through the Christian Science Board of Directors of The Mother Church. This By-Law takes effect on Dec. 15, 1908.

Agreement Required. SECT. 13. When the Christian Science Board of Directors calls a student in accordance with Article XXII, Sect. 11, of our Church Manual to the home of their Leader, Mrs. Eddy, said student shall come under a signed agreement to remain with Mrs. Eddy if she so desires, during the time specified in the Church Manual.

Incomplete Term of Service. SECT. 14. If a student who has been called to serve our Leader in accordance with Article XXII, Sect. 11, of the Church Manual leaves her before the expiration of the time therein mentioned such student shall pay to Mrs. Eddy whatsoever she may charge for what she has taught him or her during the time of such service.

Help. SECT. 15. If the author of the Christian Science textbook call on this Board for household help or a handmaid, the Board shall immediately appoint a proper member of this Church therefor, and the appointee shall go immediately in obedience to the call. "He that loveth father or mother more than me is not worthy of me." (Matt. 10:37.)

Students with Mrs. Eddy. SECT. 16. Students employed by Mrs. Eddy at her home shall not take care of their churches or attend to other affairs outside of her house.

Mrs. Eddy's Room. SECT. 17. The room in The Mother Church formerly known as "Mother's Room" shall hereafter be closed to visitors.

Pastor Emeritus to be Consulted. SECT. 18. The Mother Church shall not make a church By-Law, nor enter into a business transaction with a Christian Scientist in the employ of Rev. Mary Baker Eddy, without first consulting her on said subject and adhering strictly to her advice thereon.

o1oo (8) THE MOTHER CHURCH AND BRANCH CHURCHES

Article XXIII

Local Self-government. SECTION 1. The Mother Church of Christ, Scientist, shall assume no general official control of other churches, and it shall be controlled by none other.

Each Church of Christ, Scientist, shall have its own form of government. No conference of churches shall be held, unless it be when our churches, located in the same State, convene to confer on a statute of said State, or to confer harmoniously on individual unity and action of the churches in said State.

Titles. SECT. 2. "The First Church of Christ, Scientist," is the legal title of The Mother Church. Branch churches of The Mother Church may take the title of First Church of Christ, Scientist; Second Church of Christ, Scientist; and so on, where more than one church is established in the same place; but the article "The" must not be used before titles of branch churches, nor written on applications for membership in naming such churches.

Mother Church Unique. SECT. 3. In its relation to other Christian Science churches, in its By-Laws and self-government, The Mother Church stands alone; it occupies a position that no other church can fill. Then for a branch church to assume such position would be disastrous to Christian Science. Therefore, no Church of Christ, Scientist, shall be considered loyal that has branch churches or adopts The Mother Church's form of government, except in such cases as are specially allowed and named in this Manual.

Tenets Copyrighted. SECT. 4. Branch churches shall not write the Tenets of The Mother Church in their church books, except they give the name of their author and her permission to publish them as Tenets of The Mother Church, copyrighted in SCIENCE AND HEALTH WITH KEY TO THE SCRIPTURES.

Manual. SECT. 5. Branch churches shall not adopt, print, nor publish the Manual of The Mother Church. See Article XXXV, Sect. 1.

Organizing Churches. SECT. 6. A member of this Church who obeys its By-Laws and is a loyal exemplary Christian Scientist working in the Field, is eligible to form a church in conformity with Sect. 7 of this Article, and to have church services conducted by reading the SCRIPTURES and the Christian Science textbook. This church shall be acknowledged publicly as a Church of Christ, Scientist. Upon proper application, made in accordance with the rules of The Christian Science Publishing Society, the services of such a church may be advertised in *The Christian Science Journal*. The branch churches shall be individual, and not more than two small churches shall consolidate under one church government. If the Pastor Emeritus, Mrs. Eddy, should relinquish her place as the head or Leader of The Mother Church of Christ, Scientist, each branch church shall continue its present form of government in consonance with The Mother Church Manual.

Requirements for Organizing Branch Churches. SECT. 7. A branch church of The First Church of Christ, Scientist, Boston, Mass., shall not be organized with less than sixteen loyal Christian Scientists, four of whom are members of The Mother Church. This membership shall include at least one active practitioner whose card is published in the list of practitioners in *The Christian Science Journal*.

Privilege of Members. SECT. 8. Members in good standing with The Mother Church, who are members of the faculty, instructors, or students in any university or college, can form and conduct a Christian Science organization at such university or college, provided its rules so permit. Also members in good standing with The Mother Church, who are graduates of said university or college, may become members of the organization by application to, and by the unanimous vote of, the active members present, if the rules of the university or college so permit. When called for, a member of the Board of Lectureship may lecture for said university or college organization.

No Close Communion. SECT. 9. The Mother Church and the branch churches shall not confine their membership to the pupils of one teacher.

No Interference. SECT. 10. A member of The Mother Church may be a member of one branch Church of Christ, Scientist, or of one Christian Science society holding public services, but he shall not be a member of both a branch church and a society; neither shall he exercise supervision or control over any other church. In Christian Science each branch church shall be distinctly democratic in its government, and no individual, and no other church shall interfere with its affairs.

Teachers' and Practitioners' Offices. SECT. 11. Teachers and practitioners of Christian Science shall not have their offices or rooms in the branch churches, in the Reading Rooms, nor in rooms connected therewith.

Recognition. SECT. 12. In order to be eligible to a card in *The Christian Science Journal*, churches and societies are required to acknowledge as such all other Christian Science churches and societies advertised in said *Journal*, and to maintain toward them an attitude of Christian fellowship.

oo4o (9) GUARDIANSHIP OF CHURCH FUNDS

Article XXIV

Church Edifice a Testimonial. SECTION 1. Whereas, on March 20, 1895, the Christian Science Board of Directors, in behalf of The First Church of Christ, Scientist, Boston, Mass., presented to Rev. Mary Baker Eddy their church edifice as a Testimonial of this Church's love and gratitude, and she, with grateful acknowledgments thereof, declined to receive this munificent gift, she now understands the financial situation between the Christian Science Board of Directors and said Church to be as follows:--

Financial Situation. SECT. 2. The Christian Science Board of Directors owns the church edifices, with the land whereon they stand, legally; and the Church members own the aforesaid premises and buildings, beneficially. After the first church was built, the balance of the building funds, which remained in the hands of the Directors, belonged to the Church, and not solely to the Directors. The balance of the church building funds, which can be spared after the debts are paid, should remain on safe deposit, to be hereafter used for the benefit of this Church, as the right occasion may call for it. The following indicates the proper management of the Church funds:--

Report of Directors. SECT. 3. It shall be the duty of the Christian Science Board of Directors to have the books of the Church Treasurer audited semiannually, and to report at the annual Church meeting the amount of funds which the Church has on hand, the amount of its indebtedness and of its expenditures for the last year.

Finance Committee. SECT. 4. There shall be a Committee on Finance, which shall consist of three members of this Church in good standing. Its members shall be appointed annually by the Christian Science Board of Directors and with the consent of the Pastor Emeritus. They shall hold quarterly meetings and keep themselves thoroughly informed as to the real estate owned by this Church and the amount of funds received by the Treasurer of The Mother Church, who is individually responsible for said funds. They shall have the books of the Christian Science Board of Directors and the books of the Church Treasurer audited annually by an honest, competent accountant. The books are to be audited on May first.

Prior to paying bills against the Church, the Treasurer of this Church shall submit them all to said committee for examination. This committee shall decide thereupon by a unanimous vote, and its endorsement of the bills shall render them payable.

If it be found that the Church funds have not been properly managed, it shall be the duty of the Board of Directors and the Treasurer to be individually responsible for the performance of their several offices satisfactorily, and for the proper distribution of the funds of which they are the custodians.

God's Requirement. SECT. 5. God requires wisdom, economy, and brotherly love to characterize all the proceedings of the members of The Mother Church, The First Church of Christ, Scientist.

Provision for the Future. SECT. 6. In case of any possible future deviation from duty, the Committee on Finance shall visit the Board of Directors, and, in a Christian spirit and manner, demand that each member thereof comply with the By-Laws of the Church. If any Director fails to heed this admonition, he may be dismissed from office and the vacancy supplied by the Board.

Debt and Duty. SECT. 7. The Mother Church shall not be made legally responsible for the debts of individuals except such debts as are specified in its By-Laws. Donations from this Church shall not be made without the written consent of the Pastor Emeritus. Also important movements of the manager of the Committee on Publication shall be sanctioned by the Board of Directors and be subject to the approval of Mary Baker Eddy. (See Article I, Sect. 6.)

Emergencies. SECT. 8. The Treasurer, personally, or through the Clerk of the Church, may pay from the funds of the Church bills of immediate necessity not exceeding $200 for any one transaction, and he may keep on deposit the sum of $500 with the Clerk, as a petty cash fund, to be used by him for the payment of such bills. Such payments shall be reported, on the first of the following month, to the Board of Directors and the Committee on Finance, for their approval.

Committee on Business. SECT. 9. The Christian Science Board of Directors shall elect annually a Committee on Business, which shall consist of not less than three loyal members of The Mother Church, who shall transact promptly and efficiently such business as Mrs. Eddy, the Directors, or the Committee on Publication shall commit to it. While the members of this Committee are engaged in the transaction of the business assigned to them they shall be paid from the Church funds. Before being eligible for office the names of the persons nominated for said office shall be presented to Mrs. Eddy for her written approval.

oo3o (10) THE CHRISTIAN SCIENCE PUBLISHING SOCIETY

Article XXV

Board of Trustees. SECTION 1. The Board of Trustees, constituted by a Deed of Trust given by Rev. Mary Baker Eddy, the Pastor Emeritus of this Church, on January twenty-fifth, 1898, shall hold and manage the property therein conveyed, and conduct the business of "The Christian Science Publishing Society" on a strictly Christian basis, for the promotion of the interests of Christian Science.

Disposal of Funds. SECT. 2. The net profits of the business shall be paid over semi-annually to the Treasurer of The Mother Church. He shall hold this money subject to the order of the Christian Science Board of Directors, which is authorized to order its disposition only in accordance with the By-Laws contained in this Manual.

Vacancies in Trusteeship. SECT. 3. The Christian Science Board of Directors shall have the power to declare vacancies in said trusteeship, for such reasons as to the Board may seem expedient.

Whenever a vacancy shall occur, the Pastor Emeritus reserves the right to fill the same by appointment; but if she does not elect to exercise this right, the remaining trustees shall fill the vacancy, subject to her approval.

Editors and Manager. SECT. 4. The term of office for the editors and the manager of The Christian Science Publishing Society is one year each, dating from the time of election to the office. Incumbents who have served one year or more can be re-elected, or new officers elected, by a unanimous vote of the Christian Science Board of Directors, and the consent of the Pastor Emeritus given in her own handwriting.

Suitable Employees. SECT. 5. A person who is not accepted by the Pastor Emeritus and the Christian Science Board of Directors as suitable, shall in no manner be connected with publishing her books, nor with editing or publishing *The Christian Science Journal, Christian Science Sentinel, Der Herold der Christian Science,* nor with The Christian Science Publishing Society.

Periodicals. SECT. 6. Periodicals which shall at any time be published by The Christian Science Publishing Society, shall be copyrighted and conducted according to the provisions in the Deed of Trust relating to *The Christian Science Journal*.

Rule of Conduct. SECT. 7. No objectionable pictures shall be exhibited in the rooms where the Christian Science textbook is published or sold. No idle gossip, no slander, no mischief-making, no evil speaking shall be allowed.

Books to be Published. SECT. 8. Only the Publishing Society of The Mother Church selects, approves, and publishes the books and literature it sends forth. If Mary Baker Eddy disapproves of certain books or literature, the Society will not publish them. The Committees on Publication are in no manner connected with these functions. A book or an article of which Mrs. Eddy is the author shall not be published nor republished by this Society without her knowledge or written consent.

Removal of Cards. SECT. 9. No cards shall be removed from our periodicals without the request of the advertiser, except by a majority vote of the Christian Science Board of Directors at a meeting held for this purpose or for the examination of complaints. Members of this Church who practise other professions or pursue other vocations, shall not advertise as healers, excepting those members who are officially engaged in the work of Christian Science, and they must devote ample time for faithful practice.

oo2o (11) TEACHING CHRISTIAN SCIENCE

Article XXVI
TEACHERS

Motive in Teaching. SECTION 1. Teaching Christian Science shall not be a question of money, but of morals and religion, healing and uplifting the race.

Care of Pupils. SECT. 2. Christian Scientists who are teachers shall carefully select for pupils such only as have good past records and promising proclivities toward Christian Science. A teacher shall not assume personal control of, or attempt to dominate his pupils, but he shall hold himself morally obligated to promote their progress in the understanding of divine Principle, not only during the class term but after it, and to watch well that they prove sound in sentiment and practical in Christian Science. He shall persistently and patiently counsel his pupils in conformity with the unerring laws of God, and shall enjoin them habitually to study the Scriptures and SCIENCE AND HEALTH WITH KEY TO THE SCRIPTURES as a help thereto.

Defense against Malpractice. SECT. 3. Teachers shall instruct their pupils how to defend themselves against mental malpractice, never to return evil for evil, but to know the truth that makes free, and thus to be a law, not unto others, but to themselves.

Number of Pupils. SECT. 4. The teachers of Christian Science shall teach but one class yearly, which class shall consist of not more than thirty pupils. After 1907, the Board of Education shall have one class triennially, a Normal class not exceeding thirty pupils.

Pupil's Tuition. SECT. 5. A student's price for teaching Christian Science shall not exceed $100.00 per pupil.

Associations. SECT. 6. The associations of the pupils of loyal teachers shall convene annually. The pupils shall be guided by the BIBLE, and SCIENCE AND HEALTH, not by their teachers' personal views. Teachers shall not call their pupils together, or assemble a selected number of them, for more frequent meetings.

A Single Field of Labor. SECT. 7. A loyal teacher of Christian Science shall not teach another loyal teacher's pupil, except it be in the Board of Education. Outside of this Board each student occupies only his own field of labor. Pupils may visit each other's churches, and by invitation attend each other's associations.

Caring for Pupils of Strayed Members. SECT. 8. A loyal teacher of Christian Science may teach and receive into his association the pupils of another member of this Church who has so strayed as justly to be deemed, under the provisions of Article XII, Sect. 1, not ready to lead his pupils.

Teachers must have Certificates. SECT. 9. A member of this Church shall not teach pupils Christian Science unless he has a certificate to show that he has been taught by Mrs. Eddy or has taken a Normal Course at the Massachusetts Metaphysical College or in the Board of Education.

Such members who have not been continuously active and loyal Christian Scientists since receiving instruction as above, shall not teach Christian Science without the approval of The Christian Science Board of Directors.

Article XXVII
PUPILS

Authorized to Teach. SECTION 1. After a student's pupil has been duly authorized to be a teacher of Christian Science, or has been under the personal instruction of Mrs. Eddy, he is no longer under the jurisdiction of his former teacher.

Without Teachers. SECT. 2. Those beloved brethren whose teacher has left them, can elect an experienced Christian Scientist, who is not in charge of an association of students and who is ready for this high calling, to conduct the meetings of their association.

Basis for Teaching. SECT. 3. The teachers of the Normal class shall teach from the chapter "Recapitulation" in SCIENCE AND HEALTH WITH KEY TO THE SCRIPTURES, and from the Christian Science Platform, beginning on page 330 of the revised editions since 1902, and they shall teach nothing contrary thereto. The teachers of the Primary class shall instruct their pupils from the said chapter on "Recapitulation" only.

Church Membership. SECT. 4. Neither the Pastor Emeritus nor a member of this Church shall teach Roman Catholics Christian Science, except it be with the written consent of the authority of their Church. Choice of patients is left to the wisdom of the practitioner, and Mrs. Eddy is not to be consulted on this subject.

Class Teaching. SECT. 5. Members of The Mother Church who are authorized by its By-Laws to teach Christian Science, shall not solicit, or cause or permit others to solicit, pupils for their classes. No member of this Church shall advise against class instruction.

Teachers of Christian Science must have the necessary moral and spiritual qualifications to elucidate the Principle and rule of Christian Science, through the higher meaning of the Scriptures. "The less the teacher personally controls other minds, and the more he trusts them to the divine Truth and Love, the better it will be for both teacher and student." (Retrospection and Introspection, page 84.)

oo1o (12) BOARD OF EDUCATION

Article XXVIII
ORGANIZATION

Officers. SECTION 1. There shall be a Board of Education, under the auspices of Mary Baker Eddy, President of the Massachusetts Metaphysical College, consisting of three members, a president, vicepresident, and teacher of Christian Science. Obstetrics will not be taught.

Election. SECT. 2. The vice-president shall be elected annually by the Christian Science Board of Directors. Beginning with 1907, the teacher shall be elected every third year by said Board, and the candidate shall be subject to the approval of the Pastor Emeritus.

President not to be Consulted. SECT. 3. The President is not to be consulted by students on the question of applying for admission to this Board nor on their course or conduct. The students can confer with their teachers on subjects essential to their progress.

Presidency of College. SECT. 4. Should the President resign over her own signature or vacate her office of President of the Massachusetts Metaphysical College, a meeting of the Christian Science Board of Directors shall immediately be called, and the vicepresident of the Board of Education being found worthy, on receiving her approval shall be elected to fill the vacancy.

Article XXIX
APPLICANTS AND GRADUATES

Normal Teachers. SECTION 1. Loyal students who have been taught in a Primary class by Mrs. Eddy and have practised Christian Science healing acceptably three years, and who present such credentials as are required to verify this fact, are eligible to receive the degree of C.S.D. Qualifications. SECT. 2. Loyal Christian Scientists' pupils who so desire may apply to the Board of Education for instruction; and if they have practised Christian Science healing successfully three years and will furnish evidence of their eligibility therefor, they are eligible to enter the Normal class. All members of this class must be thorough English scholars.

Certificates. SECT. 3. Students are examined and given certificates by this Board if found qualified to receive them.

Article XXX

ACTION OF THE BOARD

Sessions. SECTION 1. The term of the Massachusetts Metaphysical College will open with the Board of Education on the first Wednesday of December. The sessions will continue not over one week. None but the teacher and members of the College class shall be present at the sessions, and no Primary classes shall be taught under the auspices of this Board.

Special Instruction. SECT. 2. Not less than two thorough lessons by a well qualified teacher shall be given to each Normal class on the subject of mental practice and *malpractice*. One student in the class shall prepare a paper on said subject that shall be read to the class, thoroughly discussed, and understood; this paper shall be given to the teacher, and he shall not allow it or a copy of it to remain, but shall destroy this paper.

Signatures. SECT. 3. The signature of the teacher and of the President of the College shall be on all certificates issued.

Remuneration and Free Scholarship. SECT. 4. Tuition of class instruction in the Board of Education shall be $100.00. The bearer of a card of free scholarship from the President, Rev. Mary Baker Eddy, shall be entitled to a free course in this department on presentation of the card to the teacher. Only the President gives free admission to classes.

Surplus Funds. SECT. 5. Any surplus funds left in the hands of the Board of Education shall be paid over annually to the Treasurer of The Mother Church.

Primary Students. SECT. 6. Students of Christian Science, duly instructed therein and with good moral records, not having the certificate of C.S.D. may enter the Normal class in the Board of Education, which will be held once in three years beginning A.D. 1907; provided their diplomas are for three consecutive years under Mrs. Eddy's daily conversation on Christian Science, or from the Massachusetts Metaphysical College Board of Education.

Healing Better than Teaching. SECT. 7. Healing the sick and the sinner with Truth demonstrates what we affirm of Christian Science, and nothing can substitute this demonstration. I recommend that each member of this Church shall strive to demonstrate by his or her practice, that Christian Science heals the sick quickly and wholly, thus proving this Science to be all that we claim for it.

If both husband and wife are found duly qualified to teach Christian Science, either one, not both, should teach yearly one class.

Not Members of The Mother Church. SECT. 8. No person shall receive instructions in Christian Science in any class in the Massachusetts Metaphysical College, nor receive the degree of C.S.B. or C.S.D., who is not a member of The First Church of Christ, Scientist, in Boston, Mass.

Only those persons who are members of this Church and possessed of the qualifications named in Sect. 9 of Article XXVI of these By-Laws shall be deemed loyal teachers of Christian Science.

oo04 (13) BOARD OF LECTURESHIP

Article XXXI
ORGANIZATION AND DUTIES

Election. SECTION 1. This Church shall maintain a Board of Lectureship, the members of which shall be elected annually on Monday preceding the Annual Meeting, subject to the approval of the Pastor Emeritus. The lecture year shall begin July 1 of each year.

Duty of Lecturers. SECT. 2. It is the duty of the Board of Lectureship to include in each lecture a true and just reply to public topics condemning Christian Science, and to bear testimony to the facts pertaining to the life of the Pastor Emeritus. Each member shall mail to the Clerk of this Church copies of his lectures before delivering them.

No Disruption of Branch Churches. SECT. 3. The Board of Lectureship is not allowed in anywise to meddle with nor to disrupt the organization of branch churches. The lecturer can invite churches within the city whither he is called to unite in their attendance on his lecture, and so make for their churches a less lecture fee; but the churches shall decide their action.

Receptions. SECT. 4. As a rule there should be no receptions nor festivities after a lecture on Christian Science, but there may occur exceptions. If there be an individual who goes to hear and deride truth, he should go away contemplating truth; and he who goes to seek truth should have the opportunity to depart in quiet *thought* on that subject.

Circuit Lecturer. SECT. 5. Upon the written request of Mrs. Eddy, The Mother Church shall appoint a Circuit Lecturer. His term of office, if approved, shall not be less than three years. He shall lecture in the United States, in Canada, in Great Britain and Ireland.

A member shall neither resign nor transfer this sacred office.

Article XXXII
CALLS FOR LECTURES

From the Directors. SECTION 1. When the need is apparent, the Christian Science Board of Directors of The Mother Church may call on any member of this Board of Lectureship to lecture at such places and at such times as the cause of Christian Science demands.

From Branch Churches. SECT. 2. The branch Churches of Christ, Scientist, may apply through their clerks to a member of this Board of Lectureship for a speaker, and one shall be assigned them by the Board.
From Societies. SECT. 3. If called for, a member of the Board may lecture for a Society.

Annual Lectures. SECT. 4. The Mother Church and the branch churches shall call on the Board of Lectureship annually for one or more lectures.

No Lectures by Readers. SECT. 5. No lecture shall be given by a Reader during his term of Readership. The duties alone of a Reader are ample.

No Wednesday Evening Lectures. SECT. 6. The Board of Lectureship shall not appoint a lecture for Wednesday evening.
Lecture Fee. SECT. 7. The lecture fee shall be left to the discretion of the lecturer.

Expenses. SECT. 8. The lecturer's traveling expenses and the cost of hall shall be paid by the church that employs him.

Exceptional Cases. SECT. 9. If a lecturer receive a call to lecture in a place where he sees there is special need, and the local church is unable to meet the expense, he is at liberty to supply that need and trust to contributions for his fee.

ooo3 (14) COMMITTEE ON PUBLICATION

Article XXXIII

In The Mother Church. SECTION 1. There shall be appointed by The Mother Church a Committee on Publication, which shall consist of one loyal Christian Scientist who lives in Boston, and he shall be manager of the Committees on Publication throughout the United States, Canada, Great Britain and Ireland. He shall be elected annually by a unanimous vote of the Christian Science Board of Directors and the consent of the Pastor Emeritus given in her own handwriting, and shall receive an annual salary, paid quarterly, of not less than four thousand dollars.

Duties. SECT. 2. It shall be the duty of the Committee on Publication to correct in a Christian manner impositions on the public in regard to Christian Science, injustices done Mrs. Eddy or members of this Church by the daily press, by periodicals or circulated literature of any sort. This Committee on Publication shall be responsible for correcting or having corrected a false newspaper article which has not been replied to by other Scientists, or which has been forwarded to this Committee for the purpose of having him reply to it. If the correction by the Committee on Publication is not promptly published by the periodical in which it is desirable that this correction shall appear, this Committee shall immediately apply for aid to the Committee on Business. Furthermore, the Committee on Publication shall read the *last proof sheet* of such an article and see that it is published according to copy; he shall circulate in large quantities the papers containing such an article, sending a copy to the Clerk of the Church. It shall also be the duty of the Committee on Publication to have published each year in a leading Boston newspaper the letter sent to the Pastor Emeritus by the Church members in annual meeting assembled. The State Committees on Publication act under the direction of this Committee on Publication.

In Branch Churches. SECT. 3. The Readers of the three largest branch churches in each State of the United States and in Canada shall annually and alternately appoint a Committee on Publication to serve in their localities. For the purposes of this ByLaw, the State of California shall be considered as though it were two States, the dividing line being the 36th parallel of latitude. Each county of Great Britain and Ireland, except as hereinafter specified, through the Readers of its three largest branch churches, shall annually and alternately appoint a Committee on Publication to serve in its locality. Each church is not necessarily confined to its own members in selecting this Committee, but if preferred, can appoint a Committee on Publication who is in good fellowship with another Church of Christ, Scientist.

This By-Law applies to all States except Massachusetts, in which the Committee on Publication is elected only by the Christian Science Board of Directors. The Committee for the counties in which London, England, is situated shall be appointed by the Christian Science Board of Directors, and he shall, in addition to his other duties, act as District Manager of the Committees on Publication for Great Britain and Ireland.

Appointment. SECT. 4. The Committees on Publication shall consist of men generally. Each State Committee shall be appointed by the First and Second Readers of the church employing said Committee. If prior to the meeting of the church for the election of officers, Mrs. Eddy shall send to the First Reader of the church the name of a candidate for its Committee on Publication, the Readers shall appoint said candidate. Or if she shall send a special request to any Committee on Publication, the request shall be carried out according to her directions.

Removal from Office. SECT. 5. If the Committee on Publication neglects to fulfil the obligations of his office according to these By-Laws, and this becomes apparent to the Christian Science Board of Directors, it shall be the duty of the Directors immediately to act upon this important matter in accordance with said By-Laws.

The Christian Science Board of Directors may notify any Church of Christ, Scientist, to remove its Committee on Publication and to appoint another Committee to fill the vacancy; and it shall be the duty of that church to comply with this request. In such cases it shall be the privilege of this Board to name the Committee if it so desires, and any Committee so named by the Board shall be elected by the branch church.

Case of Necessity. SECT. 6. If a suitable man is not obtainable for Committee on Publication, a suitable woman shall be elected. If at any time the Christian Science Board of Directors shall determine that the manager of the general Committee on Publication needs an assistant, the Board shall, with the approval of the Pastor Emeritus, appoint an assistant manager, who shall receive an adequate salary from The Mother Church.

0002 (15) CHURCH-BUILDING

Article XXXIV

Building Committee. SECTION 1. There shall be a Building Committee consisting of not less than three members, and this committee shall not be dissolved until the new church edifice is completed. This committee shall elect, dismiss, or supply a vacancy of its members by a majority vote.

Designation of Deeds. SECT. 2. All deeds of further purchases of land for The First Church of Christ, Scientist, in Boston, Mass., shall have named in them all the trusts mentioned in the deeds given by Albert Metcalf and E. Noyes Whitcomb in March, 1903; but this rule shall not apply to land purchased for any purpose other than the erection of a church edifice. Also there shall be incorporated in all such deeds the phrase, "Mary Baker Eddy's Church, The Mother Church or The First Church of Christ, Scientist, in Boston, Mass."

The Mother Church Building. SECT. 3. The edifice erected in 1894 for The First Church of Christ, Scientist, in Boston, Mass., shall neither be demolished, nor removed from the site where it was built, without the written consent of the Pastor Emeritus, Mary Baker Eddy.

ooo1 (16) CHURCH MANUAL

Article XXXV

For The Mother Church Only. SECTION 1. The Church Manual of The First Church of Christ, Scientist, in Boston, Mass., written by Mary Baker Eddy and copyrighted, is adapted to The Mother Church only. It stands alone, uniquely adapted to form the budding thought and hedge it about with divine Love. This Manual shall not be revised without the written consent of its author.

Seventy-third Edition the Authority. SECT. 2. The Board of Directors, the Committee on Bible Lessons, and the Board of Trustees shall each keep a copy of the Seventy-third Edition and of subsequent editions of the Church Manual; and if a discrepancy appears in any revised edition, these editions shall be cited as authority.

Amendment of By-Laws. SECT. 3. No new Tenet or By-Law shall be adopted, nor any Tenet or By-Law amended or annulled, without the written consent of Mary Baker Eddy, the author of our textbook, SCIENCE AND HEALTH.

End of By-Laws - Appendix omitted.

Editorials

The book is a part of my Kaleidoscope project that is designed to highlight items of significance for understanding, and for the healing, of failures that gave rise to the greatest existential challenges in modern time, for individuals, nations, and the world.

Some of the challenges are the presently ongoing financial and economic collapse that has swept across the imperial West; the terror orgies, the fascist environmentalisms and depopulation policies; the extreme nuclear-war threats against Russia and China, with boasts that such a war is winnable; and not least the critically ignored start of the next Ice Age in the 2050s, if not sooner, in which the territories outside the tropics become uninhabitable, where presently most of humanity lives. In these arenas the power of love, humanity, scientific honesty, and spiritual development have fallen by the wayside.

The following editorial "Fighting for the Truth" is designed to focus on only one aspect, a rarely known aspect on the front of the failures that are fundamental to the current crisis. The aspect focused on, in the editorial, is the despotic nature of elitism, ranging from ecclesiastical and scientific despotism, all the way to political despotism.

Fighting for the Truth

By Rolf A. F. Witzsche - 12/3/2016

The ancient scourge of ecclesiastical despotism, rather than having been healed, has been allowed to expand into the domains of science, civilization, politics, and economics. The world is in a rut. Who speaks for the truth?

Oh yes, the world is in a rut indeed on all of these fronts. Physical Science, for example, has been choked to death with the scam-doctrine of manmade global warming, or manmade climate change as it is now called, which has not one aspect of truth in it, for which upwards to 100 million people are murdered each year with starvation under the biofuels hoax. Vast amounts of agricultural resources are diverted to be burned as motor fuels for no benefits whatsoever. It takes huge areas of lands to grow the crops that become fomented into roughly 600 billion gallons of alcohol hooch, that becomes distilled to over 90 billion gallons of 99% pure alcohol, termed ethanol that is added to gasoline and burned with it, at 65% lower energy output. Nor is it pollution free. The mass-burning of this alcohol-type fuel produces a number of cariogenic air-pollutants. It is a neurotoxic, psychoactive drug. When it is burned, it produces atmospheric formaldehyde, acetaldehyde, carbon monoxide, nitrous oxides, ozone, and nearly doubles the carbon dioxide emissions when one counts the emissions from the burning of carbon fuels that are required to produce the ethanol. Almost as much energy input is required to produce the ethanol fuel than the fuel gives back. All considered, the biofuels process is a highly expensive and inefficient energy conversion process, rather than being a net energy producer. It is highly efficient, only in murdering people. The volume of agricultural resources that are diverted to be burned, would normally nourish 400 million people. In a world that has a billion people living in chronic starvation, the mass-burning of food is murdering most likely 100 million people a year, for no benefit whatsoever, with vast segments of society participating in the holocaust at the gas pump. That's the face of science despotism, which society obediently bows to. Political objectives now determine with known lies, what is scientific truth. And this is only one single aspect on a wide horizon of similar aspects, and the smallest of them in terms of consequences.

Science-despotism, if it is not reversed soon, may ultimately destroy civilization and 99% of humanity with it. The day of this happening is not far distant. A large volume of physical evidence exists that a phase shift in the solar process will likely occur in the 2050s timeframe, which will diminish the Sun's surface temperature from the present 5,800 degrees (Kelvin) to roughly 4,000 degrees, for a 75% reduced solar energy output; with which the next glaciation cycle begins, commonly termed the

Ice Age. The weakening of the solar system towards the phase shift is already in progress. The evidence is so numerous that nearly two dozen video presentations were required to present it. The presentations can be found on my website under the heading, "Cool Science for Kids to have a future in an Ice Age World." See: http://www.ice-age-ahead-iaa.ca/000/index.html

The Ice Age transition itself, promises to be a quick one, in the range of days and weeks, not the thousands of years that mainstream science is preaching. The time we have left to get ready is short, indeed, and the task is enormous, though not impossible.

To get ready, we need to built 6,000 brand new cities from scratch and have them completed and occupied in roughly 30 years. The new cities are required to enable the relocation of all the nations that a presently located outside the tropics, into the tropics, together with their agriculture, their industries, and cultural institutions. Since there is little suitable land in the tropics, the new facilities and infrastructures will have to be placed afloat across the equatorial seas.

To fail on this front amounts to committing mass-suicide. Researchers tell us that only 1-10 million people had made it through the last glaciation period alive. This is roughly what the primitive Earth can support by itself in an Ice Age world, without technological infrastructures. Without truth in science, and following that, truth in politics, culture, and economics, we are lost. Lost here means, to be people without a future.

The entire novel that these 4 parts are elements of, named Lu Mountain, which is the last of a sequence of 12 novels, is focused on the question, 'where do we go from here?' This asks, what progress must we make to have a future?

The obvious answer is that the train to the future must stop at many stations along the way. The 4 parts presented here are stations far down on the track. The first station needs to be named: Despotism Ends Here.

The break with despotism is what the story of the posters with dragons is focused on. If we cannot get to this 'station' on the track, the land beyond cannot be reached. This is what the CSD concept signifies. It was developed in the early years of the 1900s by a New England woman, named Mary Baker Eddy. She was one of the great pioneers of her time.

She had suffered a severe spinal injury in 1866, that was said to become fatal. While near death, she found herself suddenly healed, by some evidently profound spiritual reasoning. She set out to discover the science of what had made this healing possible, and in the course of it, began to teach others to heal on a scientific spiritual basis. She founded a college for this purpose, in 1883, that became known as the Massachusetts Metaphysical College. In the mean time she founded a church to perpetuate and expand her science, which she named Christian Science. As this task required all her time, she closed the college in 1889, in which she had personally taught over 4,000 students. She stopped teaching, but retained the charter for her college, and her position as its president. She reopened it 10 years later as an auxiliary to her Church in a symbolic manner, by making no provisions for teachers that would teach in the college. Instead, she set up a Board of Education that would carry on some basic teaching, but was only allowed to issue the bachelor's degree CSB. She awarded the higher degree, CSD, to those who were taught by herself personally. Here comes the crux of the matter that remained almost hidden. In the appendix to her Church Manual, she provided sample application forms that stipulated that the membership application needs the countersigner to be a person who has passed examination by the Board of Education, or one who has taken a degree at the Massachusetts Metaphysical College. With the college having no teachers provided for it, the process of taking a degree at the college remains only possible symbolically, and is only possible to be certified by the accomplishments achieved. It is as if she was saying to society that at the leading edge one is alone. No one is qualified there to stand as judge over another, render examination, and award certificates. In this case the pioneer closed the door to ecclesiastical despotism, if only in principle.

This pioneering example is important, because in our modern time science has become choked with certified despotism. The entire manmade global warming doctrine rides on the elitism of certified despotism. Truth no longer has a voice. This also means that society no longer owns its own thinking. Truth becomes politically determined and despotically imposed. It is almost impossible in modern time for society to get itself off the Global Warming Train to hell that runs in the opposite direction, away from the future, hiding the real nature of the Ice Age dynamics in order that humanity will not create itself a future, but lay itself down to die, instead. The science despotism serves the dream of mass-depopulation to preserve a feudal system of empire that has no place left in the world to actually exist. As a looting system, it has destroyed what it is feeding on.

The challenge to overcome despotism is evidently one of the greatest challenges in modern time. Mainstream astrophysics has become a club of certified liars, like a number of other departments of science from politics to economics to environmentalism. Without the despotism that chokes the world, the needed 6,000 new cities would likely be built with ease, which seems presently impossible, and definitely so under the tragically still-prevailing western financial doctrine of private imperial monetarism. The tragedy that maintains the despotism that chokes the world, especially the western world, is the disease in civilization, called elitism. Elitism and despotism are one, and this one renders humanity so small, mentally, that society now argues against its very survival. If the 6,000 new cities are not built, that enable the relocation of the nations outside the tropics, into the tropics, civilization will end when the solar system's Ice Age phase shift occurs. The despotism of elitism almost assures this. It assures this with the same certainty as the military nuclear-war-posture doctrine of Mutually Assured Destruction, ultimately assures the destruction of the world and nearly all of humanity with it. The disease of elitism in civilization, which keeps the mind small, boxed in, and artificially confined with politically motivated lies against man, will murder humanity collectively with near-perfect assurance if the disease is not healed. It is a fatal disease. It has always been that. It is a disease that demolishes the truth in the sight of man as if it didn't exist. The disease renders everything mortal, even humanity

itself. It preserves nothing. It diminishes everything, by which the most precious crumbles into dust.

The story in this book has 4 parts. The first part is focused on reaching for the pinnacle of what is good and beautiful in humanity, which is the foundation for civilization.

The second part is focused on the rights of humanity to know the truth about itself, and its freedom rooted in the truth.

The third part is focused on understanding and on acknowledging in a scientific manner the power that each one of us has as a divine human being, to uplift the world, in which the value of the human being comes to light.

The fourth part is focused on the renaissance potential that is inherent in a spiritually uplifted self-perception in society. That's what economics is all about, isn't it? Economics is built on a spiritual foundation, of Love expressed as love, meeting the human need as efficiently and as fully as possible. Economics is a process of uplifting physics into the realm of metaphysics. The whole of society presently lives by man-created resources that are enabled by scientific and technological processes, which far supersede in their creative and productive capacity what the natural world, on its primitive platform, is able to provide.

Further references for this type of 4-part progression can be found in my book, *Christian Science and Christ and Christmas*. (see updates as: http://www.ice-age-ahead-iaa.ca/christian_science/Christian_Science_and_Christ_and_Christmas.html*)*

As a highly developed spiritual humanity, we have achieved wonders on this path of scientific and spiritual development, though we have barely begun to develop this potential. In spite of the grand achievements that have been wrought from time to time throughout history, the Ice Age Challenge before us puts all of these achievements into the category of a mere beginning. I say this in comparison with the creative and productive power that is inherent in our humanity as human beings, and as the greatest gems on the planet.

The potential still exists at the present time to meet the Ice Age Challenge and to assure one-another a bright and richly livable future. However, if we continue to bow to the despotism of elitism that chokes humanity as the terribly dangerous cultural disease that it is, we will remain stuck in the easy chair under the thumb of despotism, and do nothing towards having the grand human future that we are capable of creating. In this case we cannot escape the consequences that promise to be infinitely worse than the horrors of the world wars in which society has destroyed continents and nations as a 'little' people from the throne of self-denial at the bidding of its despotic elite. We can stop this train of history and reverse direction. Russia and China have already begun this reversal, and made a significant starting contribution in a number of grand respects with their commitment to war avoidance and world development. Whether this is enough, and unfolds fast enough to inspire the world to comply with the astrophysical Ice Age Schedule, remains to be seen. The world is in the twilight in this respect. Nevertheless, in considering the speed with which the turn-around is already unfolding, inspires justified confidence that the vastly greater Ice Age Challenge will yet be met.

We will Build 6,000 New Cities in 30 Years and we will do it for free.

By Rolf A. F. Witzsche – 02/03/2016

We face a little challenge. The challenge is to build 6,000 new cities from scratch, for free, for a million people each, in 30 years. If we master the challenge we will all continue to live, and live securely and richly, in the brightest renaissance world of all times.

Should we fail ourselves, by doing nothing, we would be committing universal suicide by starvation and would murder our children with it. The stakes are that enormous, because the driving factor is the Sun.

In 30 years, our star, the Sun, will revert back to its dimmer state, its Ice Age state, that has been its natural state for 85% of the last half a million years, and probably for much longer than that. The warm period that we enjoyed for the last 12,000 years, termed the interglacial period, is an anomaly that has run its course and is in the process of ending. The normal ice-climate is about to resume. That's the nature of the universe. The clock is ticking. The evidence that the impending phase shift is near has been measured in several different ways, by satellite, by carbon-14 ratios, by ice-coring projects, by solar activity cycles, by magnetic pole drift, by coronal holes on the Sun, and by other types of astrophysical occurrences that are in progress. The evidence is vast and is imperative.

When the astrophysical phase shift occurs that becomes reflected massively in the dynamics of our Sun, potentially in the 2050s, or slightly earlier, the radiated solar energy promises to be 70% less than it is today, according to available evidence. That's when the next Ice Age begins, termed the glaciation period. It begins with the type of deeply cold climates that reflect the 70% loss in radiated solar energy.

The reduced solar intensity poses somewhat of a problem for us all, right across the world, without exceptions, especially in agriculture. The weaker Sun will render most of the regions outside the tropics rapidly uninhabitable; in possibly as short a period as a single year or less. But what of it? We, humanity, can live with that. We can compensate for the loss in solar energy by simply relocating ourselves, and all of our northern agriculture, into the tropics where the sunshine is stronger by almost the same amount, in comparison, by which the solar intensity becomes reduced. This means that we can continue to live there quite well. Agriculture will continue to be possible in an Ice Age in the tropics. In practice, it may be enhanced with artificial environments. All of this is possible. But will we do it?

It is physically possible for all people who presently reside outside the tropics or in fringe areas, to relocate themselves into the tropics. For this, 6,000 new cities will need to be built, for a million people each; together with new industries, new transportation systems, new energy systems; and above all, new infrastructures for agriculture. And all of this will need to be completed in the 30 years that we have still remaining in the current interglacial warm period that is itself, as I said before, an anomaly.

The task to build 6,000 new cities in 30 years is vast in scope and appears to be somewhat bewildering when seen in the traditional context, but as a technological project it can be relatively easily accomplished with automated, large-scale, high-temperature industrial processes, that utilize basalt as a feedstock. Basalt is a finely grained stone that melts at 1,400 degrees. With the appropriate heat-processing, basalt can be shaped into any construction product that is desired, from complete housing modules, to floating agricultural modules, floating cities, even floating forests. For the lack of suitable land in the tropics, the new civilization will have to be built largely afloat across the equatorial seas.

Can we do this? Of course we can. With large-scale automated industrial processes and infinitely available materials and energy, almost any construction project can be accomplished with ease. With these, we have the power to live as kings on the Earth in the richest imaginable civilization, should we care to do so. But will we?

We have demonstrated in the past that we are more inclined to do nothing, and remain glued to the Easy Chair, such as we presently are. If we continue this path, we won't have the power to live past the solar phase shift. In this case we would commit ourselves and our children to an agonizing death by starvation, as a default response. A middle path between the two options is not possible.

That a middle path is not possible is glaringly evident by the recognized fact that only 1 to 10 million people worldwide, have emerged from the last glaciation period alive. This small number of people is evidently all that the natural environment can support in an Ice Age world, unaided by technological infrastructures. By us all remaining stuck in the Easy Chair we would commit the greatest-ever genocide in the history of the world. The genocide may even lead to the extinction of humanity in the potential wars over the scraps of food that may remain from before the phase shift. We have the nukes now in our pocket to turn the world into hell, and we have the agendas in place to use them for far lesser reasons.

Against the background of our potential self-extinction, either by nuclear war or by facing the Ice Age unprepared, I like to suggest that we will choose the option that enables us to live, as unlikely as this choice presently seems. However, I would suggest that we, as human beings, are not that small to remain forever asleep, so that we will write ourselves a ticket for a living world of our creating, by doing whatever it takes. If the task requires us to build 6,000 new cities for a million people each, and produce them for one-another for free in order to enable the relocation of all people out of the areas on the world that become uninhabitable in 30 years' time when the next Ice Age begins, then this is precisely what we will do.
 Especially the youth of today will inspire itself to get this done. Its existence is at stake. It will see the great challenge as a new type of dance, and will have fun with implementing the solutions.

It is always immensely joyful to see great projects unfolding. Against this background the building of

6,000 new cities will become an exciting opportunity, an opportunity to experience the creative power that is inherent in the human being, but has been bottled up for far too long by the murderous effects of imperial cultural warfare projects. The 6,000 new cities will be built. They will be built as a demonstration of human freedom, and they will be built in a mode of celebration. The celebration of this self-claimed freedom in society will break the chains of empire. This step is a step that has not been accomplished in 5,000 years to date, but will likely be accomplished now, and with ease.

The Ice Age Challenge is not a challenge to repair or transform the present systems of finance, economics, politics, and science, as such transformations to repair the dead horses are not really possible. Instead of the defective old phenomena being repaired, they will be simply overlaid and left to fade into the dust of history. Those who say that the 6,000 new cities cannot be built in 30 years, are perfectly correct, in the context of the present platforms. But when the dust settles, a whole new world will come into view. That's totally possible, because when humanity raises itself up to higher levels, above the dust, where it deals with reality and with power, the impossible becomes achieved routinely. If this was not so, we would have no hope.

The barrier here, that still stands in the way, is self-evidently small-minded thinking. Society doesn't allow itself to become real. Those who say that "hell will freeze over" before society will acknowledge the cosmic Ice Age dynamics, are perfectly correct. Mainstream science has become a frozen wasteland where nothing moves in real terms. It cannot be reformed or be repaired, but it can be left behind in the dust when honest science develops and inspires humanity.

At the present time, there is no willingness that I have encountered anywhere, to face the real challenge. People argue, as they have been told to argue, that the Earth is flat, that the Ice Age is not real, that the Sun is an invariable constant. And so, mainstream science has blinded itself with the dust it loves, always afraid of manmade global warming that's just a dream. It's the song that everybody sings. The song refrains, "There is no such thing as cosmic dynamics that cause Ice Ages. The Sun is a rock-solid fire in the sky heated from within by nuclear fusion that has never varied and never will. There is no need to get out of the Easy Chair. No change will occur, not for a thousand years. Cyclical variations of the orbit of the Earth causes the Ice Ages."

Every excuse that is possible, is being dredged up in these songs that people sing while they remain glued to their easy chair, nicely asleep in dense ignorance. But that's a deadly song. It is the song that empire demands to be sung for this very reason. The song of ignorance is not the song of life. I would suggest that a human song be sung, the song of science, joy, and power, in which the task ahead is hailed as but a small and interesting chore.

The evidence is plain when one looks from a higher-level platform where reality is the key factor. In order to be able to survive past the 2050s, humanity will have to create itself 6,000 entirely new cities for a million people each, in order to be able to relocate the majority of its populations into the tropics, from outside the tropics that become largely uninhabitable when the Ice Age phase shift happens in the 2050s or slightly earlier. That's the human song.

The human song is to put away the childish notions that this is a big task, and to start dealing with 'power' – real power – to get it done. The outcome is inevitable, if we step up to the power-level that we are able to. A new industrial revolution and cultural renaissance beyond anything we have seen in the past, will be the outcome. The alternative is not a human song.

It is never a human song that inspires humanity to let itself collapse to less than 10 million people, as we had coming out of the last Ice Age. The nature of our humanity is to move ahead towards evermore powerful modes of living. That's the human song. The tasks involved are easy tasks, as easily accomplished, as singing.

Most of the current agriculture is endangered by the solar phase-shift that causes a massive climate-change in the 2050s when the solar surface temperature drops from 5,800 K to slightly less than

4,000 K, which adds up to a 70% reduction in radiated energy, with which the next Ice Age begins. The down-ramping towards this phase shift has been in progress for more than 3,000 years already. The phase shift nearly happened in the 1600s. The solar dynamics were in an extremely precarious state then, with almost no sunspots visible for nearly an entire century. Fortunately for us, the phase shift was prevented before it happened. It was prevented by the latest of the Dansgaard Oeschger 'pulses' that were discovered in the Greenland ice cores as huge cyclical phenomena that occur in intervals of roughly 1470 years. That this 'pulse' that rescued us in the late 1600s has ended in the late 1990s is well documented in measurements by the Ulysses spacecraft, in the diminishing solar activity cycles, and in the numerous fringe-effect types of climate changes, of which the California drought is an example of many similar aspects. Here is where the dynamics of our humanity has a chance to come to light.

The California freshwater crisis, for example, cannot be resolved as a single issue in isolation by itself, because it is an effect of a much larger dynamics that is cosmic in scale. The all-overriding challenge is to respond to the large cosmic dynamics. The challenge thereby becomes a challenge to raise the economic platform far beyond the greatest historic achievements and the most daring economic concepts; even to raise it up by orders of magnitude above the imperial platform that rules the world today, especially in the sciences; that also rules in Russia, India, and China to some degree. The challenge is a challenge to get out of the dust bowls of empire. The science of astrophysics is largely grounded in the dust bowl today. People argue for what is not real, and deny what is. They are committing genocide over it by default, and become murderers of their children.

The only, actually critical danger that we face in the larger context, is that society is not willing to face reality and its imperative. As I said, humanity still remains trapped in the dead landscape of science-denial that the cultural warfare-projects from H. G. Wells onward, have accomplished, where reality is turned upside down and most of it is deemed not to exist. Society loves its familiar science-fairy-tales, a type of mental pabulum that it is spoon-fed for the diminishing effects the pabulum has, that are intended. Dreaming, asleep in the Easy Chair, requires no efforts in society. Dreaming with the eyes closed requires no scientific honesty. In this sense, the easy chair is free. Society is free there, free of the nagging demands impelled by the truth. Society is free there to lay itself down to die. Yes, the Easy Chair is that deadly. It is deadly, because it prevents the physically needed responses to reality.

We face deadly consequences on this path where reality is not acknowledged. It is a path that is paved by simple-minded faith without understanding. The simple-minded path has become a crowded highway in modern time; a highway of mysticism; dogmatism; learned ignorance, especially in astrophysics; and of politically correct beliefs, such as in entropy, in the Big Bang cosmology, in manmade climate change, and in the dream of the endless interglacial in which the Sun remains a constant forever without variance. Mainstream science owns much of this highway. The most trusted elite is parading thereon. Almost everyone is found there. Except there is no truth beaming on the horizon. The horizon is a dark black cloud, dim with inhumanity, war, terror, and even thermonuclear war terror; and the foreground is correspondingly bleak, desolate, lifeless.

The people on the highway say to me, Rolf you are nuts, the Ice Age is still a thousand years distant and is so slow in coming that we have eons of time to react to it. My answer to them is, look at the evidence; the phase shift towards the end of the interglacial has already begun; the evidence is monumental; it is everywhere; stop the dreaming; you are committing your children to death by starvation with your dreaming. You are forcing the collapse of civilization with your dreaming. Civilization is doomed to collapse when society finds so little value in its humanity, in itself, that it actively prevents itself to have a future, and more than that, denies itself the joy that unfolds with the realized power of a great renaissance that flows from the acknowledged value of humanity.

The physical phase of the Ice Age Challenge is in its pre-stages of unfolding. There are enormously-large physical cosmic dynamics in progress already on a wide front, as the interglacial is coming to its end.

Based on research that began more than 25 years ago, I have produced over the last few years, dozens of videos to illustrate the nature of the ever-changing cosmic dynamics that we cannot evade by any means that we possess, or will likely ever possess.

I have also illustrated the consequences and the opportunities involved. The consequences are of a type that we cannot prevent, but they have corresponding imperatives attached for us if we care to continue to live past the end of the interglacial, that is in the final stages of ending. The building of 6,000 new cities is an example of the imperatives. This is an imperative in the strongest sense, because if this imperative is not met, history ends, because no one will write it. It the new cities are not built that enable the bulk of humanity to relocate itself into the tropics before their territories become uninhabitable, the ensuing human tragedy is unimaginable. The imperatives are of a type that we cannot get away from, but are able to respond to.

Ironically, it is here where humanity's greatest imperatives are located, which is also where society is failing itself. Humanity has created miraculous space exploration capabilities, and made amazing discoveries with it, but ironically it looks at its discoveries with its eyes closed and its mind barred, obediently, as it has been trained to do for more than a century already. This self-imposed blindness has created a crisis that society has been trained not to see. The cosmic dynamics are on the move. Big moving 'events' are plainly visible on the horizon. Only the corresponding humanist movement that should have begun decades ago, has not yet begun. By its failing to move, society has deprived itself of the brightest and richest renaissance it could have, and could have had already decades ago, and would have had if the Ice Age imperatives would have been acknowledged when they were first raised in the late 1960s, which where smothered under imperial directive with the manmade-global-warming science-travesty in 1974. That is where the crisis is rooted that is now upon us, which society is unwilling to recognize as a crisis, obediently. Without this blocking factor, we would be living in an entirely different world.

Without the blocking factor standing in the way, the

Ice Age Challenge would be met with the power of the great renaissance that we are fully capable of creating as human beings. On this higher-level platform that a renaissance is built on, where the value of the human being is supremely understood and acknowledged, it becomes possible to built the needed 6,000 new cities in 30 years to protect and honour this greatest element of value that exists in the world, which is the human being.

We are presently far from this stage. We are at the dismal stage where the human being is devalued to near zero. We hail the destruction of humanity as a panacea. We have created immense war machines at enormous cost for the intention of the destruction of one-another. We have even placed nuclear war on the agenda, with the capability to eradicate the human presence completely in the space of a lunch break. Why would anyone even dream about building 6,000 new cities in this context of madness, for humanity to live past the interglacial and to have a bright future under the worst conditions?

Inversely, if the supreme value of the human being was understood and acknowledged in society, the needed 6,000 new cities would be built as a matter of fact in a renaissance perspective, and the orgies of murder and destruction would fall by the wayside. This is what it means for humanity to move with power. The power of humanity lies within its nature as the greatest manifest of life on the Earth, with the greatest creative capacity that has ever been developed on this planet. When this becomes acknowledged the building of the needed 6,000 new cities is assured. Physically, the task can be accomplished with ease with the material and energy resources that we have on hand in great abundance, and with the already established technologies that have been sitting on the shelf unused for half a century already, which society presently refuses to utilize, obediently.

As I said before, the 6,000 new cities can be created quite easily with very-large-scale automated high-temperature industrial processes that utilize basalt as feedstock and nuclear power to drive the process. The power of large-scale automated industrial processes is so great, and so efficient, that the resulting products can be given away for free, and will be once the process begins. In fact, nothing less than universal free housing would be sufficiently large in developed humanist power to meet the Ice Age Challenge and its timeline.

Free high-quality housing produced by humanity for itself - as an investment into itself - is the minimal foundation for a powerful civilization to stand that meets the continuing challenge.

Human cultural development begins with constructing high-quality housing for all across the board. Homelessness, slum living, rent slavery, and so on, add up to the greatest waste imaginable, of the most precious resource that a human society has, which is itself by its human potential. The Ice Age Challenge literally demands us to snap ourselves out of the currently accepted self-diminishing mode that is a deadly trap laid out intentionally by the system of empire that pervades almost all levels of society.

The development of truthful science, that takes us out of the dust-bowl of imperial constricted science, is the natural product of intentionally advanced cultural progress in a human society. For this progression, high-quality housing stands as a critical foundation. The connection between this foundation and superstructure that becomes a renaissance in which life is assured, is obvious. In the light of this obvious fact, it is gradually becoming recognized in society that the trap of homelessness and slum living is an patently imperial objective that has been promoted for more than a century from Thomas Malthus onward, and continues to be promoted to the present day, but which can be overturned with relative ease. The Ice Age Challenge, literally impels us to snap us out of the 'heart-choking' traps of the systems of empire as fast as possible.
Do I hear objections from anyone, against the need for this humanist freedom, universally? No I don't. The truth of it is knowable. What is truth, is knowable, and evermore of it is gradually becoming known. We are in a crisis today, because the process has not been driven forward fast enough. While much has been accomplished, vastly more needs yet to be accomplished, even as the remaining time in the interglacial is fast running out.

The Ice Age Challenge, therefore, has a critical role to play in every respect. If it was acknowledged fully today, it would not only snap us out of the present housing crisis, but would also snap us out of a similar trap that inhibits agricultural development. We should be far past the stage of the present near-total reliance on outdoor agriculture.

Indoor agriculture with controlled lighting, controlled environments, controlled CO_2 concentrations, and so on, can raise the power of agriculture-efficiency by several orders of magnitude. While floating agriculture strung across the equatorial seas might have a role to play in the coming Ice Age world, in the traditional outdoor manner, the result is presently uncertain. It is inherently far-more efficient to place agriculture right from the start into scientifically optimized and technologically maintained environments, enhanced with artificial lighting.

When the mere doubling of the CO_2 concentration in commercial greenhouse operations increases plant growth by 50%, we are looking at potentials before us that we have barely begun to exploit. The choking effect of monetarism has so far prevented any significant progress on this front, just as it has on every other critical front. It appears therefore, that nothing short of the critical development needs that come in the form of a single package with the Ice Age Challenge, will break the chokehold that monetarism presently lays upon civilization.

Of course, housing and agriculture are not the only arenas where an honest response to the Ice Age Challenge would impel a highly-powerful new renaissance. The renaissance process would also fulfill the age-old hope to have a world without war. This hoped-for potential now stands before us as an achievable reality.

This is not utopian thinking. A world without war is an immanently realizable potential, a potential that we have at hand as a part of the package of the end of empire and its derivatives that the Ice Age imperative demands us to achieve, and to achieve fast. Empire equals war. War equals destruction. War and Empire IS One. Both aspects are a singularity that is a package and ends as a package. It is not possible to end war without ending the system of empire that imposes war. There is no compromising way possible that would separate the two aspects of the singularity. And this recognition even includes

thermonuclear war. Thermonuclear war is not an exception. It is a part of the package. The package cannot be transformed to make it liveable. This has been attempted for 70 years. Instead of being more secure, the world is more precarious than it had been during the Cuban Missile Crisis. The imperial game has been so intensified that the time from launch to impact has been reduced to roughly 5 minutes in the most critical areas. It is not possible that humanity can ride out this increasing storm indefinitely.

The only imperative that can end the immense danger of a sudden or accidental thermonuclear war, is the Ice Age Imperative. This imperative has this power, because the imperative, by its very nature, impels the greatest renaissance ever imagined. No other imperative in the world today impels this kind of renaissance power that, for example, can produce 6,000 new cities in 30 years as a means to assure the continued living of humanity. No other imperative has the power to raise the humanist platform so high that the present empire-bound world falls by the wayside, including its wars, its greed-based system that injects impotence into economics, and its terror orgies and psychological-warfare projects that induce the self-regression of society across the entire front of civilization, including in the sciences. The Ice Age Imperative impels a higher-level platform in human self-perception, and corresponding actions, than has ever been attained before on the universal stage. With this imperative the Ice Age Challenge will be met. It is a case of uplifted truthfulness where the power to move with the truth is rooted, which is the only real power that we ultimately have. On any lesser stage, society does not operate with power.

The present economic landscape that is mired in depravity of numerous kinds, is devoid of the power to accomplish anything significant for the advance of humanity. It is a dead-end street. At the higher level of social action that is termed morality, we stand above depravity, but insufficiently developed to recognize our potential. Morality is an insecure stage. For example, the mass-depopulation of the planet is deemed an acceptable moral goal for the benefit of the planet. We murder 100 million people with starvation every year with the biofuels hoax, which is deemed so moral that almost everyone participates at the gas pump. We burn food products converted into biofuels that would normally nourish 400 million people. We take food from the needy and burn it for no benefits at all. This is still considered moral by the vast masses who participate in the process. It takes a scientific revolution in the self-perception of society as human beings, to raise itself to higher ground where the holocausts of today are impossible to contemplate and to carry out, where higher-level concerns move us.

That this stage is still termed utopia is evident by the fact that the Ice Age challenge is nowhere on the agenda, and the scientific imperative in terms of physical evidence that would place onto the agenda is being denied and hidden, even by the most-moral-seeming elements in society. Major scientific developments are needed here, in terms of society's self-perception as human beings, before the Ice Age Challenge is being acknowledged. The field is presently quite empty with nothing much moving there. And even if the level of 'utopia' was achieved, which the poet Friedrich Schiller described as "The Sublime," we would still have to raise ourselves up one more level to develop the power to meet the physical component of the Ice Age Challenge.

It is not a small matter for society to rebuilt its entire world in less than 30 years. To get the power developed to do this, the self-perception on society has to be raised to a level to the near divine. The Ice Age Challenge demands the near impossible. On any lesser stage the building of 6,000 new cities is deemed impossible and will thereby be prevented.

In order to meet the great challenge that is imposed on us by the Ice Age schedule, like building 6,000 new cities with new agriculture on technologically created 'land' afloat on the tropical seas, nothing less will suffice than the uplift of humanity itself, from its self-perception as an impotent biological creature, to its native divinity as the greatest treasure that exists on Earth with the creative capacity to built up the means to preserve its life in the normal Ice Age world and beyond.

With this up-lifted self-perception in society, as a treasure of great value and with near-divine power, the nuclear-war challenge falls by the wayside, together with the driving force behind it, which the system of empire is. History has shown that nothing

short of the gigantic renaissance the Ice Age Challenge impels us to realize as an existential challenge, will be sufficient to break the deadening effect of the past and leave it behind us in the dust of forgotten history.

This means that the Ice Age Challenge is not a challenge that becomes critical in 30 years. It is critical now. The creative uplift in society is needed now. Last year, the year 2015, was the 70th anniversary year of the folly of nuclear war and of the politics that drive the folly. We find ourselves standing more deeply mired today in this anti-human madness than ever before, with the engines of war standing ready to cause the extinction of humanity in the space of a lunch-break.

The question needs to be asked, why is the self-eradication of humanity still on the policy agenda? Why are we so small in our own sight to stand impotent against this existential threat as if this impotence was a facet of human character? The nuclear-war danger is real. Somebody sneezing the wrong way may trigger that hell that no one will not be affected by, and very few, if any, will be able to survive.

Why do we find so little value in ourselves, that we stand so small today as human beings, that we have allowed this trap to be sprung on us and have not found the power to resist it and overturn it in all those 70 years? Here the twin-factor of the Ice Age Challenge may help us.

The Ice Age Challenge stands not primarily as an existential challenge that must be met for humanity to survive. It stands more profoundly as a challenge to our creative spirit to develop the physical resources to protect our living in a manner that the ice age dynamics have no effect on it. It is the renaissance component that gives the Ice Age Challenge the potential to succeed. This is the factor that the nuclear-war front, by itself, doesn't have, or does not exist on any other front, economic, financial, or otherwise. The Ice Age Challenge is also the only challenge that is not artificial. We need to be honest here, and not play games.

The divine nature of our humanity requires a commitment to love on a universal platform as a matter of principle. That's a tall task. To fail, is rich with tragedy. Society has failed itself miserably in the past. In 1914 all of Europe became each other's enemy on piles of lies accepted that vanquished love, and as a consequence the nations that defied the principle of universal love, butchered each other to death. Those who died, fell not by some divine vengeance, but by their smallness in which they gave the boot to love, singing boisterously, "fuck you!" And so they 'fucked' themselves and perished in the ditches and so forth. By the time that dust had settled, more than 50 million lay dead. The next war was 'fought' on the same platform, singing "fuck you!" instead of reaching out a helping hand to one another. This time 100 million people perished, and a continent lay in ruins. The tragedy occurred outside the framework of the divine principle, Love, expressed universally.

Society still sings the "fuck you" song in economics, politics, nuclear-war terrorism, and so on, though it is slowly beginning to move away from these. While the pace is significant, it is far from sufficient to enable the building of 6,000 new cities in 30 years, and all that goes with it, with the whole world joining hands in universal love for one-another's humanity and its divine nature as the tallest aspect of life on the planet. To fail here is universally deadly.

The Ice Age Challenge is real. We cannot avoid it. Nor should we deny it. We should embrace it, because it really does inspire us to up-lift the quality in ourselves as human beings that we already have, and raise it up to the divine level of ourselves where we deal with power, with creative power, to become paragons of productive energy with which to create for ourselves the kind of richer world that naturally meets all the Ice Age challenges of the future, and this even now while time is running out fast.

The actual science for the timing of the Ice Age dynamics is rather simple. The evidence is plain. The dynamics and the evidence are all easily recognizable by the principle involved, as in the case of the Meno Dialog in Plato's Republic, where a slave boy was challenged to double the area of a square, and did so with the full understanding of the principle that provided the solution. The boy tackled a difficult challenge that is almost impossible to meet outside the complex domain of seeing with the mind's eye. The Ice Age Science Challenge is of the same quality,

and is no more difficult to resolve in the complex domain. It is simple enough for a child to meet the challenge in the mind when the evidence is honestly presented, because the fundamental science aspects are largely rudimental in nature. Many of the physical aspects are taught in the schools in physics classes, though only to the point that political imperatives allow this. I have presented the Ice Age Science Challenge extensively under the heading "Cools Science for Kids to have a Future in an Ice Age World."

Politically, the Ice Age Challenge is much more difficult. The challenge that is involved with facing the coming Ice Age was recognized strongly enough in the early 1970s, so that the scientific community suggested that a world-conference should be convened to discuss the steps that would be needed for humanity to protect itself against it. The request was accepted, but the agenda was turned upside down. The conference was convened in 1974 in Bucharest. But instead of the Ice Age Challenge being discussed, the Manmade Global Warming doctrine was dished up in the tour of force that still continues.

It continues without evidence. It continues without scientific truthfulness. It continues in a landscape of lies where real evidence is being denied or twisted to serve the desired doctrine.

The tour of force of science perversion has prostituted mainstream science to sing the doctrine's song. The cause of the truth has suffered a huge defeat in the process. In this defeat society has lost much of what has remained of its divine love for its humanity. Humanity has become trapped into numerous economically and socially destructive processes, such as the genocidal mass-burning of food under the biofuels project. These are deeply sub-human, and sub-divine processes that render humanity essentially, functionally dead. A society that evades or denies the Ice Age Challenge, is functionally dead, which is the outcome that the evasion assures. Towards this end, by policy for depopulation, the primary target of the modern imperial science inversion project is the science of astrophysics itself, with a special focus on inverting the science aspects of the Ice Age dynamics.

It has been recognized in imperial circles as far back

as the days of H. G. Wells that the advance of scientific progress, and the thereby advancing self-recognition in society of the value and power of the human being, dooms the reign of the oligarchic system that is devoid of power, the system of empire, the system that produces nothing of value for society, but gains its wealth by clever acts of stealing, or looting, or profiteering, or whatever the term is used to hide the empty system behind a fake gilded facade.

The oligarchic 'master', H. G. Wells, had delivered his anti-science message to the world's oligarchy with his novel, The Time Machine, saying to its elite, in essence, that if you don't suppress science, the Morlochs will eat you up for breakfast. It appears that Wells' message was understood. The whole world now prays on its knees to the god of science inversion. It obediently fears the totally impossible manmade global warming and is inflicting huge sacrifices on itself, all in honour of the god of terror that the science inversion has established. The terror of the nuclear bomb, and of economic destruction with the doctrine of monetarist derivatives gambling, are all nothing more than scientifically honoured inversions of the truth that society devotes its living to. Thus, in honour of the god of inverted science, the real science of the Ice Age Dynamics remains hidden, and with it the divine quality of universal love in humanity, remains blocked, whereby the future of humanity is in doubt.

The Ice Age Challenge demands that all of these debilitating mythologies be set aside. Few are willing to do this. Some say to me that I am hopelessly optimistic if I expect this to happen. They say that especially the breakout from science inversion will never happen, it won't be allowed to happen, so that the 6,000 new cities will never be built that would enable the relocation of the nations from outside the tropics, into the tropics, before their territory becomes uninhabitable by the solar phase shift that ushers in the next Ice Age.

While such arguments are typically arguments of self-accepted impotence in mind, or as Friedrich Schiller has lamented in his time that the great moments in history have found society a 'little' people, the fact remains that ultimately, when all the dust is brushed away, humanity is inherently the most-powerful species of life that exists on this planet, or has ever existed here.

We are the children of a near infinite mind in terms of our reach both into the vast expanse of physical space and into the future that has not yet been. In this context, the Ice Age dynamics are precisely knowable by our developed capability to know the truth. Whatever is truth, is knowable because it is true, contrary to all doctrines, beliefs, conjectures, dreams, and tales that are told. The age of knowing the truth is dawning upon the world.

It is tempting to assume that this dawning is impelled by the Ice Age Challenge before us, making its claim, because the challenge is so great that it impels us to reach higher than we have ever reached in discovering what we are capable of as creators and producers of new worlds, which we are as human beings. While this may be so, no direct evidence for it exists, in the light that the Ice Age dynamics are largely denied.

It is also tempting to assume that the already unfolding Ice Age dynamics have a physical effect on humanity's neurological processes by electric induction produced by solar cosmic-ray flux that is rapidly increasing. Historic evidence does exist that this might be so.

In addition, political movements are afoot that are aimed at war-prevention on a platform of universal, cooperative, economic development towards meeting the common aims of all mankind. The BRICS system is one of the starting gates on this path, as is China's "One Belt - One Road" universal economic development commitment on a continental scale. With these seeds already sprouting, the greatest of all projects, the Ice Age Development Imperative, may yet have a chance of succeeding. And so, I would say that the 6,000 new cities that this imperative demands, will be built, and will be built on time, together with everything else that goes with it that honours the value of the human being as a matter of principle.

The recognition of the Ice Age Challenge itself is already a testament of the greatness of the human being over physical, mental, and political limitations. We, humanity, have reached deep into space and have seen the evidence of the Ice Age dynamics unfolding there, as for example the Ulysses satellite has that has revealed many critical dynamic aspects that were unknown before, and would have remained unknown without the massive space exploration that has been undertaken.

While society largely doesn't care to acknowledge the discoveries that have been made, but clings to its old comfortable dreams, the great discoveries remain nevertheless a part of our heritage now as human beings, and as a part of our understanding of the science of the Ice Age Dynamics that affects us more deeply than any other occurrence in the history of civilization.

Most of the discovered evidence for the Ice Age dynamics is new. Some great discoveries have also been made on the surface of the Earth itself, such as with the ice coring projects in Greenland and Antarctica. Most of the critical discoveries related to the Ice Age dynamics were made after the year 2000. The tendency to ignore these discoveries, which have already drifted into the background as unimportant, is synonymous with humanity ignoring itself and becoming 'little' people again as the masters have bid us all to become. Nevertheless, the numerous Ice Age dynamics related discoveries will likely not vanish from the horizon, since the dawning of the truth has already begun on a wide and horizontal platform where it may not be stoppable by any means that the masters of empire possess.

And so, I must repeat that some progress has been made that gives one hope that the 6,000 new cities will be built that enable humanity as a whole to meet and master the Ice Age Challenge, which the whole of humanity owns as a challenge. With all this in mind, I would like to suggest that the entire Ice Age Imperative will be accomplished, with floating agriculture attached to intercontinental floating bridges, serviced by floating cities, and powered by cosmic, plasma-electric energy.

While we haven't even begun to acknowledge the imperative, I would like to predict that all faucets of it will be accomplished and be completed on time, and this not merely as a means to continue to live, and live securely, because when agriculture fails, no one eats. Without food, people die. The nations of the world then become refugees in a dying world with no refuge to go to. I would like to predict that this reactive, as impelling as it may seem, is not sufficient as a power to drive the imperative. The imperative will ultimately be carried by something much more esoteric, than a reactive response. It will be carried by the human soul, because it is the human thing to do. It will be carried not as a challenge, but as an opportunity to experience the greater depth of our humanity. It will be carried to completion and beyond by the divine universal love for our precious humanity that we hold in the heart as the most valued gem in the universe of life, which it is a joy to behold and to experience and to be a part of, without which we would likely not even exist.

This doesn't mean that we all don't have a critical role to play in the process of uplifting ourselves and our world and one-another. We all have an immensely critical role to play in all that. Nothing happens by itself. The human future is something great that will never be, unless it is created by us to be.

Unlike the moth or the elephant, who flourish and fade according as the wind blows, and some become extinct by it, we, humanity are the masters of our destiny. That's what the 6,000 new cities are all about. We have the power to change the world for our living, before the cosmic forces act on it and change it for us. We have the power to dance circles around the cosmic impositions, and come out richer every time we dance. That's something to keep in mind to direct our future with, as we wish it to be. That's exciting, isn't it? It is certainly worth all the steps that are required to get us there.
The end